CONSTRUCTION PROJECT
ADMINISTRATION IN PRACTICE

CONSTRUCTION PROJECT ADMINISTRATION IN PRACTICE

A. A. KWAKYE MSc, F.R.I.C.S, A.G.I.S, M.C.I.O.B.

Addison Wesley Longman
Addison Wesley Longman Limited
Edinburgh Gate, Harlow
Essex CM20 2JE, England
and associated companies throughout the world

Co-published with The Chartered Institute of Building through
Englemere Services Limited
Englemere, Kings Ride, Ascot
Berkshire SL5 8BJ, England

First published 1997

British Library Cataloguing in Publication Data
A catalogue entry for this title is available from the British Library

ISBN 0-582-29461-4

Set by 32 in (10/12pt EhrardtMT)
Produced through Longman Malaysia, PP

CONTENTS

PREFACE

The production of construction products is a risky, complex and lengthy process. It involves several specialist participants with diverse skills and passes through several distinct but interrelated phases. For reasons of simplicity, these distinct phases may be categorised as pre-contract and post-contract, with several divisions within each phase. This book sets out to provide, in one concise volume, the administration of procurement of the construction project. Procurement is a word only recently encountered in the context of construction work and, therefore, has a different meaning to many people. In this volume, the term is considered to include the total process of procuring construction project from the client's initial idea through to his or her occupation of the completed facility. Therefore, those aspects of design and production and the related problems of briefing, documentation and organisation which affect the cost of construction projects have been tackled. In the process, functions of all participants and their contractual relationships from inception to completion have also been discussed. In addition, the problems associated with the intricate nature of the construction process have been identified and various approaches adopted to resolve them have been explained in understandable terms.

This volume has been written with the aim of providing Higher National Diploma/Certificate Students, students on degree courses in Construction, Construction Management, Quantity Surveying, Building Surveying, Architecture; and also students on Quantity Surveying, Building Surveying and Construction Management courses with an appreciation and groundwork in all aspects of the construction process. Experienced practitioners will also find this book an interesting and useful source of reference.

The main themes developed are fourfold. Firstly, the peculiar characteristics of the construction industry within its framework as the construction process evolves; secondly, the pre-contract activities starting from client's development decision to the contractor's appointment; thirdly, the post-contract activities which involves the site production process; and finally, the current alternative arrangements for construction procurement.

The early chapters examine what goes on in the pre-contract phase of a traditional procurement system, highlighting the various stages in the process from the examination of how a construction project is formulated, designed and documented. The later chapters seek to demonstrate the post-contract activities of a traditional procurement system.

The last two chapters describe the integrated procurement system and discuss why it is gaining increased acceptance and changing the construction scene.

Finally, since the construction process has been examined from the administrative stand point, reference to any particular standard form of contract has been avoided. Therefore, where necessary, students are advised to familiarise themselves with the various standard forms of contract listed in the text. Also, further reading sections have been provided for those readers who may wish to study particular topics of the book in greater depth.

I should like to express my gratitude to all those who have helped in writing this book.

My thanks to Ben Beke BSc, MSc, ARICS, AGIS, MCIOB; Gavin Cole; Raymond Floyd MCIOB; Robert Fraser BSc; and John Zeffertt BSc, ARICS, MCIOB for their constructive advice and assistance on many aspects of the book, which have taken a great deal of their valuable time.

Grateful thanks are also due to Florette McCammon for her patience and splendid work in typing and amending the numerous drafts.

Finally, a heartfelt thank you to my wife Tina for her continued assistance and support.

A. A. Kwakye

CHAPTER 1

GENERAL NATURE OF CONSTRUCTION

1.1 INTRODUCTION

In most countries around the world, the production of construction products takes place within an all-important industry, the construction industry. It is a complex and comprehensive industry which remains largely low technology, low skill and labour intensive but is responsible for the procurement of investment products for promoting production process or for direct consumption. The construction industry can be best described as a collection of industries because generally a completed building is composed of an assembly of building materials/components and equipment produced by other industries. Moreover, construction has unique characteristics deriving largely from the physical nature of the product and a group of activities interconnected by the nature of their products, technologies and institutional setting. Hence, construction activities cover a wide range of loosely integrated groups of participants/organisations and sub-markets over a wide geographical area. Together, these contribute to the production, alteration, refurbishment, maintenance and repair of fixed capital products which, in turn, contribute to the well-being and functioning of a modern national economy.

The construction industry plays an important role in the national economy. In 1993, for example, it employed 14 per cent of the UK workforce. In addition, the value of output in the whole industry in 1993 was £46.3 billion, which represented nearly 10 per cent of the gross domestic product. This also represented nearly 50 per cent of the gross fixed capital formation of the UK and holds a similar position in most advanced and developing economies of the world. Traditionally, the industry has been used by governments for national construction/reconstruction and as the monetary value of projects it undertakes is high, the industry's major client has tended to be the central government. As a result, the government plays a very significant role in the affairs of the industry, and its action on national and local level determine the level of demand for construction activities, cost of production, supply conditions and so forth. Furthermore, owing to its size, any improvement in the efficiency of the production process and/or in the quality of the finished product has the potential for large cost savings and energy conservation.

1.2 THE CONSTRUCTION INDUSTRY

The construction industry is basically a manufacturing industry and has particular characteristics which require consideration in order to explain its specialisation, complexity and organisational structure. The reasons for the particular mode of operations in the UK construction industry, for example, can be best explained as the result of its size, diversity of operations in wide geographical areas, organisation of production process, price determination procedures, number of specialist firms required and various sources of equipment and material/components needed for production. The peculiar characteristics by which it differs from other manufacturing industries (e.g. the car industry) may be summarised as follows:

1. *Fragmented industry:* The construction industry is fragmented in terms of the number of institutions representing construction professionals, the number of trade associations representing tradespeople and specialisation within the industry. Hence, the industry is weak in its consultative process and, for this reason, can neither lobby effectively nor let its voice be heard in high places.
2. *Irregular employment:* The industry is characterised as having a larger than average level of unemployment as a result of its inability to match capacity with demand at any point in time. The industry may work under capacity at one moment, which means full employment for all workers in it, but the next moment (i.e. when the national economy is in recession), over-capacity returns to the industry and, in its wake, workers face redundancy.
3. *Reliance on casual labour:* Employment of casual labour and sub-contracting instead of full-time employed operatives is a common feature of the industry. Moreover, directly employed workers tend not to be employed for long periods of time and some enter and leave the industry several times.
4. *Lack of investment:* Lack of building investment in fixed capital assets and reliance on sub-contracting and plant hire is the practice of building contractors. Plant ownership requires heavy investment in fixed assets but hiring does not and, hence, hiring provides contractors with a flexible capital that can be invested in the money market when required.
5. *Unpredictable workload:* The amount of work available to the industry and its firms over a period of time is difficult to predict. This makes it impossible to forecast the training needs, the level of investment and the level and value of each firm's output.
6. *Complex structure:* The construction industry has a complex structure in terms of the different establishments involved in design, engineering, surveying, contracting, plant hire and materials production/supply. All these make the industry susceptible to disputes, delays, avoided responsibilities and missed opportunities for innovation.
7. *Separation of functions:* Unlike other manufacturing industries, the design and production functions in the construction process are separate and, therefore, in the construction industry production begins only when the design process has ended. This is prone to lead to design without concern for buildability or

production economies and perpetuation of costly mistakes from one project to another.

8. *Lack of unity of approach:* The contracting arrangement tends to cause and sustain a *'them and us'* attitude. It is the client who often pays for this attitude as the project team tend to fight among themselves instead of tackling the problems posed by the project in unity for the client's benefit. Thus, projects often fail to meet user requirements and delays and cost overruns are endemic in the construction industry.
9. *Dependency on several industries:* The construction industry relies heavily on a number of other construction-related industries for the production and supply of a variety of prefabricated and semi-finished materials/components. This makes the industry vulnerable to the shortages and inefficient production methods of those industries.
10. *Long production cycle:* There is usually a long gestation period between inception and production phases of a construction project, and also the production time for a major project is measured in years rather than days or weeks. For this reason, sudden increases in the demand for construction products cannot be met as all the pre-construction activities will have to be undertaken for any new project proposed. Hence, the supply of construction products are inelastic, even in the long term.
11. *Transient organisation:* Production organisation is transient and each new project requires fresh management and participants from a number of disciplines and establishments to undertake a sequence of tasks which, to a degree at least, vary from project to project.
12. *Mobile operatives and equipment:* The workplace and production process need to be organised continuously with each new construction project and, hence, operatives, equipment and tools have to be moved from site to site.
13. *Operations in an ever-changing environment:* Production is highly susceptible to the uncontrollable and often unfavourable conditions imposed by the weather, and this hampers productivity and thereby increases the cost of production.
14. *Multi-party industry:* The construction industry is a multi-party industry with cash flowing from client to main contractor and then down the sub-contract chain. For this reason, those at the lower end of the chain experience any financial problems that those at the top end may endure.

1.2.1 Structure of construction industry

Construction is a fragmented industry in terms of the nature of work undertaken (Building or Civil Engineering), the technologies it uses, its clients (private and public sector) and the large range of firms/companies (professional and contracting) involved and, hence, the construction industry has a complex structure. In simple terms, the industry may be classified and structured to reflect the method of organising the construction process. Traditionally, the construction process is undertaken by two main groups - consultants and contractors – both working on

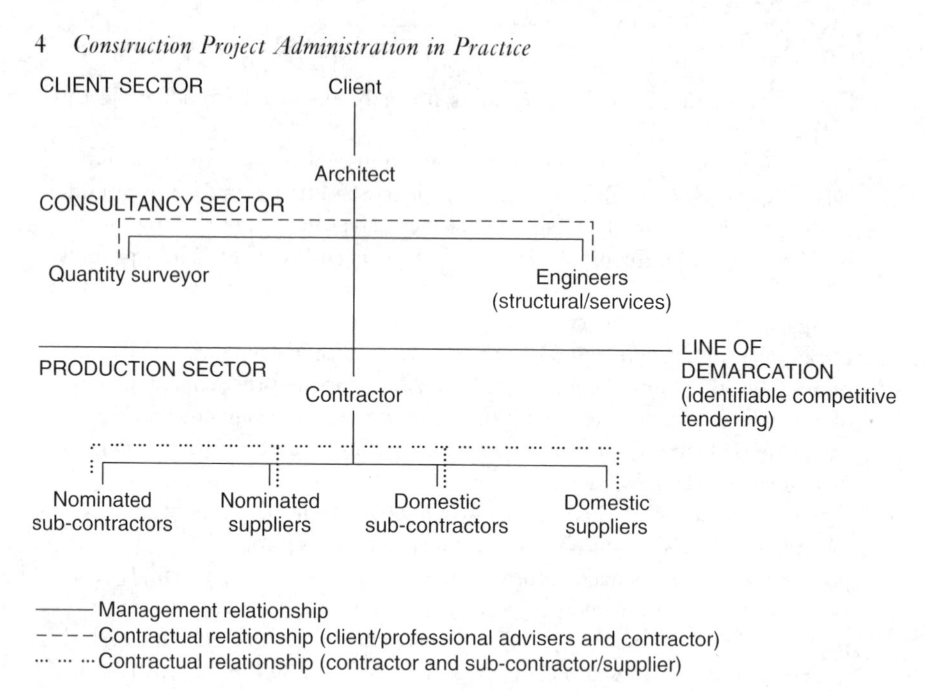

Fig. 1.1 Typical separation of design and production functions

behalf of the client (see Figure 1.1). There are also sub-contractors and suppliers who assist the contractor in various ways in the production sector.

1. *Consultants:* The consultants are architects, engineers (structural/services) and quantity surveyors who are construction professionals with diverse skills and, hence, offer design and management services for a fee.
2. *Contractors:* The contractors, specialist sub-contractors and suppliers are essentially commercial companies who supply materials/components and carry out the construction production for profit.

Conventionally, production is separated from design of the project; however, there are alternative arrangements whereby design and production can be integrated, as in Design and Build projects (see Figure 1.2 and Chapter 14).

1.2.2 Capacity of construction industry

Capacity of the construction industry may be described as a concept used by the government when it is considering the magnitude of capital expenditure projects that can be conveniently undertaken by the industry at a given time. As construction is not the leading sector of the national economy and tends to follow the peaks and troughs of other sectors, it is extremely difficult to match capacity of the industry with demand at any given time. As government is the major client of the industry, the industry can be said to be working over or under capacity, depending on the amount of public sector works that

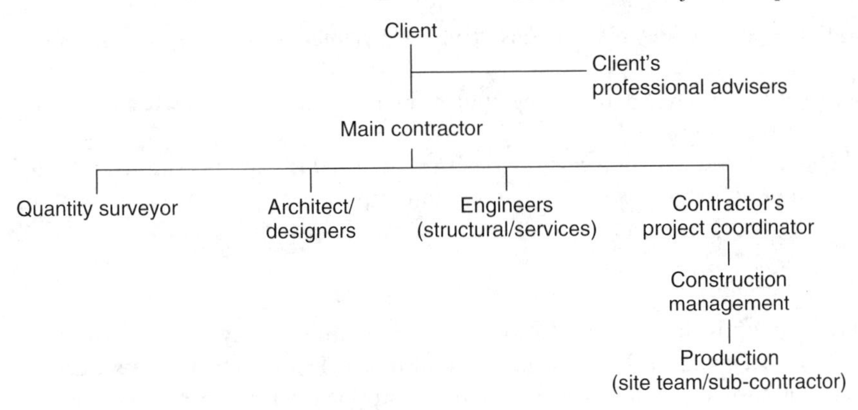

Fig. 1.2 Integrated design and build arrangement

government places on it. Therefore, capacity varies over time and also responds to demand over time.

Generally, the main resources at the construction industry disposal are human (management and operatives), materials and plant/equipment and, hence, a shortage of any of them can have a drastic effect on the industry's capacity in response to demand. Nevertheless, short-term measures can be adopted to counteract any sudden increase in demand for construction activities. Such measures include the following:

- Use of existing stocks of materials and components, unemployed and under-utilised operatives, plant and construction establishments.
- Switching of labour resources from other industries.
- Increasing productivity either by shift working or extended working hours.
- Simplification of operation to enable the use of unskilled operatives (e.g. use of dry lining in lieu of wet plaster by skilled operatives).

While in the case of operatives, materials and plant, the above measures can be effected to meet the demand for increased capacity in the short-term, there is no short-term solution for the shortage of construction managers as they cannot be trained overnight. Therefore, long-term measures, which involve educational and training systems, need to be instituted to equip personnel to undertake management roles in the construction industry. This is of utmost importance as the ability of the industry to take on more capital projects at any time depends on the availability of resources and the efficient organisation of the construction process. A long-term policy for the industry is therefore essential; but when the flow of work to the industry is less than the capacity available, a number of consequences follow, including:

- Redundancies of construction professionals and operatives due to staff reductions and/or total closure of professional firms and construction companies.
- Submission of uneconomically low tender prices by contractors, which affects profitability.

- Extremely keen fee bids by consultants, which affect profitability and quality of service.
- A cut in education and training budgets by firms and companies as a result of reduction in profits.
- Research and development suffer as a result of reduced profits and the need to trim down overheads in order to remain competitive.

1.2.3 Productivity

Productivity is the amount of goods or services produced by a productive factor in a unit of time. In the construction industry, high productivity is essential, especially when the industry, as a whole, is working below capacity. However, many difficulties are experienced in arriving at comparative figures for productivity in the industry because of the heterogeneous nature of construction work and the widespread under-recording of output and employment. Nevertheless, from available statistics, it is possible to estimate the value of construction output per man-week or man-month over the industry as a whole. It is also possible to measure productivity in the industry in terms of real cost per unit output. However, comparison of productivity on different projects and between operatives is much more difficult. Variations in output are also experienced because the day-to-day performance of operatives is influenced by the following factors.

1. *Motivation, skill and state of health of operatives:* Highly motivated skilled and healthy operatives are very productive.
2. *Difficulty and complexity of work:* Difficult and complex work slows down the rate of operatives' output.
3. *Location of Work (indoors, outdoors, higher up or below ground):* Operatives are exposed to great risk and unfavourable effects of the weather when executing operation outdoor, at a height or below ground and, hence, their productivity falls.
4. *Method of construction adopted:* Production methods are determined by the contractor's site management but operatives' performance depends on the efficiency of the method devised.
5. *Familiarity with the work:* There is always a learning curve associated with unfamiliar work and, generally, this results in a loss of production.
6. *Degree of effectiveness of plant:* Selection of plant for production is management's responsibility but operatives' performance is partly determined by the quality of plant selected for production.
7. *Degree of quality of finished work:* Where the quality of the required craft ability is high, operatives spend a certain amount of production time on the achievement of the required standard.
8. *Efficiency of planning and allocating work:* When management plans and allocates the work sections efficiently, operatives are able to perform productively in accordance with the planned allocation.

9. *Weather conditions:* Adverse effects of the weather (e.g. too hot, cold, wet) slows down operatives' performance owing to the uncomfortable conditions under which they work.

Productivity can be increased by increasing capital investment (plant/machinery), by increasing the skill of the operatives (education/training) and greater efficiency in the use of existing technologies. While increased productivity produces many benefits, this cannot be achieved without incurring costs such as those incurred in research and development, education and training. It can also be said that while the level of productivity in the construction industry does not compare favourably with that of the manufacturing industry, significant productivity gains have been recorded recently as a result of the following:

1. *Increased use of mechanised plant and tools:* The employment of improved mechanised plant and tools in construction work reduces the amount of slow physical energy required in the execution of some operations (e.g. lifting operations).
2. *Use of prefabricated material/components:* The result of industrialised construction processes has reduced site production of building components to a minimum and, hence, contractors are able to procure factory-produced building components for use on construction projects without any difficulties.
3. *Application of dimensional coordination:* The use of this technique facilitates the incorporation of prefabricated material/components into prepared openings in the building fabric without the many problems associated with this method of construction.
4. *An efficient utilisation of resources:* Improved planning and programming techniques are currently widely in use and this promotes efficient use of resources (both on and off site) on construction projects.
5. *Use of standard materials/components:* To increase productivity and reduce production time, designers have resorted to the specification of off-the-shelf standard materials/components for use on construction projects.
6. *Increased technical knowledge and management skills:* Due to pressure from clients for faster construction programmes, stricter financial control and so on, the construction industry has devised various management techniques and skills in order to cope with the ever changing construction demand.

The above are examples of some of the measures adopted for increased productivity; however, the high productivity that the above bring should not be at the expense of employment. Generally, it would not be good on a nationwide basis if high productivity on construction sites leads to a substantial unemployment elsewhere. Similarly, high productivity should not attract a high overtime payment for work executed by operatives who perform below their standard level of output. Statistically, marginal productivity falls as the working day prolongs.

1.2.4 Effects of government influence

Government influence on the construction industry has been, in part, the important contribution the industry makes towards the national economy and, in part, the way it is affected (in the short and long terms) by government economic measures. Generally, the industry remains dependent upon a wider economic stability, and if it is to flourish it requires that governments manage their respective economies efficiently to ensure solvency of national reserves, to maintain an acceptable level of employment, to maintain growth and to control inflation. But while managing the economy, governments may adopt economic measures that adversely affect the activities of the construction industry in particular (due to the way in which the industry interacts with the factors that shape a national economy). Also, from the susceptibility of the industry to government action, it can be said that the level of domestic construction work is ultimately determined by government economic policy. Governments may adopt some of the following measures to correct any imbalance in the economy and this action may, at the same time, have a detrimental effect on the construction industry.

1. *Fiscal policies:* Government employs fiscal policies to influence aggregate demand in the economy. By this practice, the government seeks to balance taxation and expenditure, and this bears heavily on the industry as follows:
   - Increased income tax reduces net income and, hence, purchasing power and this results in a fall in demand for commodities/services.
   - Tax on commodities (e.g. high mortgage interest charges) raises their prices and, hence, reduces the demand for them.

2. *Monetary policies:* Government uses monetary policies to regulate the quantity of credit that is available from the financial institutions. By this action, credit is squeezed and/or the level of interest rates is raised. The construction industry is particularly sensitive to both measures as follows:
   - A credit squeeze adversely affects construction companies who depend on short-term finance for their business operations.
   - An increase in interest rates raises the cost of projects, which causes postponement or cancellation of schemes not yet started.

The converse of the above government measures is that tax changes which put more cash in consumers' pockets (e.g. through lower income tax, tax concessions for owner occupiers and so on) eventually increase the demand for construction activities. Similarly, the lower the interest rates, the ease and lower cost at which credit can be obtained. While the industry suffers if the economy is weak, government economic measures are not attempts to regulate the economy using the construction industry; rather, it is the character of the industry which makes it prone to credit squeeze, control on demand and capital expenditure by the government. Nevertheless, government remains vital to construction activities and it can directly affect a construction workload by:

1. *Increased capital expenditure:* Financing of more public construction projects improves the trading position of the industry.

2. *Reduction of capital expenditure:* The reductions of government capital expenditure affect the construction industry because this results in a reduction of construction activities in the long term, in some sectors of the industry.
3. *Influencing the general level of demand:* This action determines the purchasing power and willingness of clients to proceed with their programme of construction investment.

1.3 CONSTRUCTION PRODUCT

Construction products are fixed capital goods which emerge as a result of building and civil engineering activities within the framework of the construction industry. Their emergence alter the configuration of the built environment and they either add to the stocks of houses, factories, office buildings and so on, or replace those that have been demolished. In economic terms, they are either produced for direct use (e.g. houses), for aiding the production process (e.g. factories) or for both (e.g. schools). Being a capital product, they are produced for long life in order to generate the necessary returns (in monetary and non-monetary terms) to cover their cost of production.

Construction product – like other durable manufactured products (e.g. cars, furniture and the like) – requires a combination of factors (land, capital, management) in its production. A plot of land/site is needed for locating and producing the construction project. Capital funds are required on a long- and short-term basis to finance plant/equipment, materials and labour power. Management is also needed to plan, organise and coordinate the production process. However, besides these similarities, the construction product has inherent characteristics which require consideration. These include the following:

1. *Peculiar product features:* An average construction product is large in physical terms and, hence, bulky and heavy (materials in particular) and very demanding on resources.
2. *Fixed to land:* The permanent anchorage to land makes the foundation or substructure a very substantial part of the product in terms of resources, cost, engineering and expertise.
3. *Durable product:* The product is durable and this accounts for the existing stock of buildings being very large in relation to commercial production.
4. *Price indicator:* The price of the existing stock of structures determines the price of new stock, and small fluctuations in the demand for the existing stock have large repercussions on the price of new stock.
5. *High priced:* The construction product demands heavy capital investment; hence, it is expensive to purchase/build and, for many clients, constitutes one of the major investments they may undertake.
6. *Production:* The construction product is assembled at the location where the ultimate consumption is to take place. Moreover, its production is sometimes hampered by adverse weather conditions

7. *Uniqueness of product:* With few exceptions, the product is unique in relation to design, location, price, engineering and production problems. As it is usually designed to order, this has tended to emphasise its aesthetic individuality and, for this reason, mass production techniques cannot be developed and utilised at a reduced cost.
8. *Cost of production:* The cost of producing a particular product is established after either negotiation or competitive tendering prior to production, and this makes its production a risky business.
9. *Specificity of order:* As the product is to order, it requires a high degree of specificity including detailed drawings, specification and procurement methods. Contract agreement and, at times, production methods, are drawn for each unit to be produced.
10. *Diverse technological requirement:* As construction products do vary in size and complexity, each calls for a different technological input in its production.
11. *Purchase/sale transaction:* Purchase/sale transactions for construction products are lengthy, complicated and often require the use of the expertise of lawyers and estate agents. This complexity attracts additional expenses to the already expensive transaction.
12. *Diverse Interest:* Interest in construction product may be diverse and proof of ownership can be quite a difficult, lengthy and costly process.

1.3.1 Demand for construction product

The demand for the construction product can be direct as well as derived. Direct demand occurs when the product is required for direct consumption. The type of construction products which fall in this category include houses, swimming pools, cinemas, car parks, educational structures, churches and hospital buildings. The demand for the construction product, when described as derived demand, expresses product for production process (i.e. product wanted because of its contribution to the production of other consumer goods and services). Construction products that correspond to this description include factories, warehouses, roads, railways and office buildings.

The demand for construction products can also be classified as either individual or collective. Individual demand arises where any individual or family feels the need for a product, has got the means and will to pay to acquire it. Collective demand, on the other hand, is a demand which arises out of a public body's desire to provide a product for collective individuals to use and has the resources and is willing to pay for the completed product.

Be the demand direct, derived, individual or collective, it is necessary to distinguish, at any given time, the demand for construction products in general from demand for construction activity. Generally, the demand for construction products would be met by the supply of existing stock of buildings. However, demand for construction activity arises when the current stock of structures is unable to meet demand and clients see the need for more buildings (new or refurbished existing stock) at profit. As the demand for construction activity regulates the need to supply more construction products, clients need to forecast this demand at some stage prior to any construction activity. The main components considered in this type of forecasting model are analysis of population growth, age distribution, sex structure, local income levels, spending patterns, immigration and availability of finance. Such a study assists clients to identify the factors that govern the demand conditions for construction products and thereby act accordingly.

1.3.2 Supply of construction product

Generally, consumer demand regulates the supply of construction products. Normally, however, the supply of construction products are unable to respond immediately to changes in demand owing to the time lag associated with the following phases of the construction process:

- Land identification and acquisition
- Design
- Funding arrangement
- Design approval
- Production and commissioning.

Normally, clients decide to build in response to an existing or anticipated demand only if they are quite certain that, on completion, their development will continue to meet the demand and, in one way or another, will be paid for. Moreover, although it is the client's development decision that eventually responds to the demand for construction product, the organisation of the production is carried out by contractors whose performance depends on resources at their disposal at any moment in time. The number and/or combination of resources employed on any construction project depend on the size of the contractor's organisation, management and operational policy, availability of resources, what is being produced, speed of production and the form of construction adopted. However, a change in the combination and/or intensity in the use of available resources (e.g. more plant utilisation, overtime and shift working) may speed up production and thus increase the supply of construction products expediently.

1.4 CLASSIFICATION OF COST

The terms *construction cost* and *cost of construction* refer to different concepts and since a considerable confusion exists about them, their clarification is important. Both terms have developed a unique identity.

1.4.1 Construction cost

Construction cost represents the amount of money a client will have to pay a building contractor for executing the construction works. Initially the contractor's tender price will give an indication of this amount. Generally, the extent of the construction cost will depend on the size, type, form, location, complexity, level of specification, tendering climate, predicted inflation, risks and procurement method adopted. However, the initial tender price may be subject to change due to one or more of the following factors:

- Increased cost as a result of increase in price of material/components, plant and human resources procurement during the execution of the project.
- Financial adjustment to the project as a result of variations authorised by the client and/or his/her professional advisers during the progress of the works.
- Claim for direct loss and/or expense submitted by the contractor due to disruptive events in the course of the production process for which the client is contractually responsible.
- Indirect cost of development which affects the production programme and for which neither the client nor the contractor is contractually responsible. For instance, an unusually bad weather which gives rise to an extension of time and which, in turn, causes the client to lose rental income or use of the product.
- Indirect cost of legal expenses incurred as a result of resolving contractual disputes in the court of law or by arbitration.

1.4.2 Cost of construction

The cost of construction can be described as the cost the contractor incurs in producing the client's building. In short, this is a contractor's production cost and basically is composed of the following:

1. *Labour cost:* Labour cost comprises the cost of employing site operatives. In the UK, this cost comprises wages, national insurance schemes, holiday schemes, pensions and so on.
2. *Materials cost:* Materials cost comprises the cost of materials purchased from builders' merchants or supplied by nominated suppliers. This cost is usually paid by means of monthly credit accounts and payment is made at the end of the month following that in which the materials were delivered to site.
3. *Plant cost:* Plant cost is the cost of hired/purchased item of plant brought onto site and used in the production process. This cost includes extra charge for bringing an item of plant to site, maintaining and removing it from site. Similar to material purchases, the hire cost is paid for by means of monthly credit account.
4. *Overheads:* Each of the above elements of cost (labour, plant and material) can be related to a specific item of site production. Any element of cost which is not confined to an individual item of work may be conveniently described as a project overhead. Project overheads cover financial matters which relate to the entire project and may be considered under two subheadings as follows:

   - *Site overhead:* This covers the cost of providing site management and equipment required for the execution of a construction project. This cost is not incurred unless a project is undertaken and it is therefore known as a direct cost. Site overhead includes the cost of maintaining site supervisory staff, offices, temporary roads, security, mess facilities, power and the like.

- *Head office overhead:* This element of cost is sometimes described as an establishment charge and includes all costs the building contractor incurs in running the construction business. Like the site overhead, it cannot be related directly to any individual item of work. Moreover, it is not confined to any particular project. Head office overhead is incurred once the contractor commences trading and, therefore, has no bearing on whether or not he/she undertakes construction work.

1.4.3 Total project cost

The term *total project cost* is the client's total expenditure on the project, which is the sum total of factors such as land costs, construction cost, interest charges and fees.

1.5 HEALTH AND SAFETY IN CONSTRUCTION

Construction is generally a dangerous operation and, for this reason, the construction industry has a poor health and safety record. Moreover, the atmosphere on a construction site is characterised as 'tough' and the work 'difficult' in nature and, hence, requires the employment of physically fit and strong operatives. But the increased pace of production, reduced size of gangs, increased use of sub-contractors and casual employment preclude the selection of healthy operatives for every stage of construction production. As a result of the foregoing, and of other factors such as unsafe designs, adoption of unsafe production methods and human errors, accidents occasionally occur and operatives get injured and/or killed. The accident rate in construction is very high and has proportionally a higher number of deaths than any other industrial sector. While this state of affair affects not only operatives but also members of the general public, it is often found after an accident that someone has been incompetent, negligent and/or careless due to one or more of the following causes:

- The overall project plan excludes concern for health and safety of operatives and all those associated with the construction project.
- The people connected with planning the project lacked knowledge of current health and safety legislations.
- The project was not designed to reflect health and safety aspects of the various production operations.
- There were special hazardous operations not mentioned in the contract documentation.
- Owing to a lack of effective safety supervision and monitoring during production, operatives were inclined:

- not to think about the safety aspects and, hence, not act safely;
- not to pay sufficient attention in order to avoid an accident;
- to behave with a general indifference to the safety aspects of the project;
- simply not to take care of themselves, other operatives or the work.

To address the poor safety issue, the UK government has over the years initiated measures aimed at reducing accidents and high fatality rate on construction sites. The most popular of these measures is the Health and Safety at Work, etc., Act 1974 which applies to all industries and offers a means by which unsafe activities on construction sites are discovered and the scope for dangerous ones to persist can be reduced. This Act and many others on health and safety (e.g. the Factories Act 1961 and the various regulations under its wings) have been supplemented by the Health and Safety Executive and the European Union's proposed Construction (Design and Management) Regulations, popularly known as *CDM*. The CDM Regulations came into force in March 1995 and are designed to enshrine the European Temporary and Mobile Construction Sites Directive in UK law. These proposed regulations impose statutory duties on all project participants and, in addition, carry criminal sanctions if contravened.

Essentially the CDM Regulations require all participants to construction to give more time and thought to planning a construction project safely for enhanced safety during production. They establish a safety management network at all stages of a construction project. In addition to imposing safety obligations on all project participants, they create two new safety duty holders: the *Planning Supervisor* and the *Principal Contractor*. The planning supervisor is not necessarily an individual; rather it is more a role and function to be legally discharged by a corporate entity with multi-disciplinary support enabling every aspect of construction to be properly understood and considered. The principal contractor, on the other hand, is an organisation that takes over and develops the health and safety plan and coordinates the activities of all contractors to ensure compliance with the health and safety law.

In all, the main aims of CDM Regulations are getting safety measures in construction right at source through:

- Selection of those with the competence and resources.
- A strategic approach to health and safety in project design, planning, preparation and production.
- Effective management and coordination of health and safety throughout construction projects.
- Restoration and improvement of overall safety management for a construction site.
- Preparation of a health and safety plan which serves as the foundation of safety management for a construction site, providing safety information for building contractors when they tender for construction work and addressing safety issues that arise during production.
- Production of a health and safety file which contains information on the project for use by anyone carrying out maintenance work in the future.

Duties imposed by CDM Regulations

As noted above, the CDM Regulations assign statutory duties on clients, designers, safety supervisors, principal contractors and contractors. These duties may be summarised as follows:

Duties of clients The client's key duties are, as far as is reasonably practicable, to:

- Check, select and appoint competent persons (design team, planning supervisor and principal contractor) to discharge those duties of health and safety on his or her construction project.
- Provide information with regard to health and safety to enable the appointees to plan the construction project safety.
- Check and be satisfied that the appointees have and will allocate sufficient resources in the interest of health and safety when working on the project.
- Provide appointees with information relevant to health and safety on the project.
- Ensure that a satisfactory health and safety plan is available before production commences on site.
- Keep, maintain and make available for inspection the health and safety file, after the project is completed.

Duties of designer The designer's key duties are, as far as is reasonable practicable, to:

- Keep clients informed of their duties under the CDM Regulations.
- Design to avoid foreseeable risks to all those involved in the construction or affected by it.
- Reduce risk at source wherever practicable.
- Provide information about aspects of the construction project (structure/ materials) which might affect the health and safety for inclusion in the health and safety plan.
- Reduce and control risks which cannot be tackled at source by instituting appropriate health and safety measures.
- Ensure that the design includes adequate information on health and safety.
- Cooperate with the planning supervisor and any other designer working on the project.

Duties of planning supervisor The duties of the planning supervisor are summarised as the coordination of the health and safety aspects of the project. They include:

- Notification of planned construction project to local Health and Safety Executive (HSE).
- Coordination of designers' efforts and ensuring that they comply with their duties - in particular, the avoidance or reduction of risk.
- Passing on relevant information from the client to the designers and/or incorporating the information in the health and safety plan.

- Preparing a health and safety plan before the appointment of the principal contractor.
- Agreeing any changes to the health and safety plan necessitated by design development and any other factors with the principal contractor.
- Advising the client on the competence of project participants and their ability to allocate resources to the health and safety plan before the commencement of production.
- Preparing, maintaining and delivering (on completion of project) the health and safety file to the client.

Duties of principal contractor The principal contractor is responsible for safety on site and his/her key duties include:

- Developing the outline health and safety plan into a working management document and seeing to its implementation.
- Selecting and arranging for competent and sufficiently resourced contractors to carry out the production when sub-contracted.
- Ensuring that contractors cooperate on safety issues during production and have all the relevant information about risks on the construction site.
- Selecting the control measures required to be employed to protect against site risks.
- Agreeing with the planning supervisor any fundamental changes to the health and safety plan.
- Monitoring, coordinating and supervising the health and safety aspects of the construction production.
- Ensuring that operatives are adequately consulted, informed, instructed and trained on health and safety issues on site.
- Providing information to all designers and passing information to the planning supervisor for the health and safety file.

Duties of contractors Contractors in general have duties to play their part in the successful management of health and safety during the production phase of the project. Their key duties include:

- Providing information about risks to health and safety arising from their work and protective measures required for the health and safety plan.
- Managing their work to comply with rules contained in the health and safety plan and directions from the principal contractor.
- Keeping the principal contractor informed of injuries, dangerous occurrences and ill health and also providing information for the health and safety file.
- Providing health and safety information to their operatives.

In summary

Although the CDM Regulations will be expensive to implement and monitor, they reinforce the need to coordinate and manage health and safety aspects of construction production from preliminary design to commissioning. This statutory duty, therefore, calls for early client's and designer's active involvement in the establishment of procedures for hazard identification and risk assessment. As it is their responsibility to specify the materials/components as well as the construction programme, they should be in a position to identify and incorporate into the contract documentation any hazards arising from their decisions. It also places a duty on all project participants to provide information on health and safety during design and production phases of the project. Furthermore, it introduces a coordinating mechanisation whereby, during the design/planning phase, the planning supervisor collects information and prepares a health and safety plan presented in a format that satisfies the need of all project participants.

1.6 STAGES OF THE CONSTRUCTION PROCESS

The construction process under a traditional system where the design and production functions are separated is characterised as a sequential approach and follows the order of briefing, scheme design, documentation, tendering, construction and commissioning of a building. Each phase is completed and approved before proceeding to the next (see Figure 1.3).

These interrelated processes associated with the construction process are constrained by time, resources and performance. They also involve a wide range of participants with practical and professional skills. The processes in each of the phases in Figure 1.3 are shown in Figure 3.2 and Table 3.1 (following the RIBA plan of work).

As the activities in Table 1.1 and Figure 1.4 show, various project participants perform different functions in each phase of the construction process. Every effort is made by the participants to solve all problems posed by the construction project and endeavour to achieve a product that meets the client's expectations.

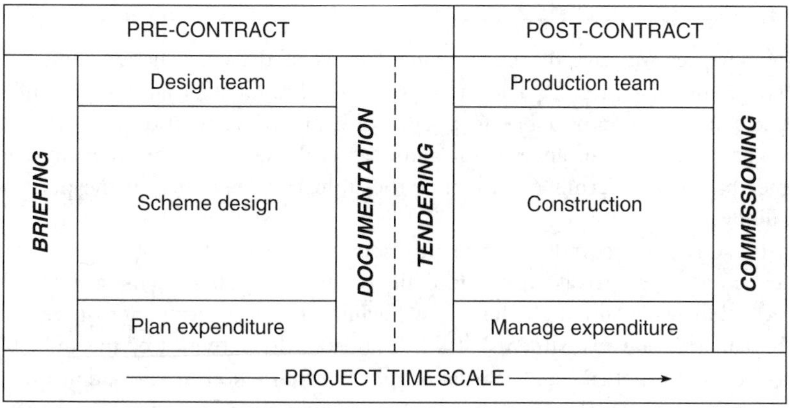

Fig. 1.3 Sequential activities

Table 1.1 Stages of construction activity

Conception/briefing	The building owner having decided to develop a piece of land appoints an architect. He or she briefs the architect of the user requirement and cost limit (i.e. the amount of capital available for the project).
Design	The appointed architect explores the feasibility of the building owner's proposal, then carries out the process of designing a building that meets the client's requirements in terms of accommodation, cost, quality and time.
Documentation	The building owner's professional advisers select the type of contract that suits the project in terms of cost, time and general market conditions. A tender document is prepared to aid selection of a suitable builder and the execution of the construction works.
Tendering and estimating	Completed documents are sent to selected contractors to enable them to submit competitive bids for the project. In most cases, the contractor who submits the lowest tender price is awarded the contract.
Construction	The selected builder undertakes the construction of the building to the shape, size and quality as depicted on the architect's drawings and specification.
Commissioning	The completed building is handed over to the building owner and the architect ensures that the building and its services perform to the owner's expectations. The architect gives the owner general guidance on maintenance and as-built drawings for buildings, drainage and service installations.

SUMMARY

In this chapter we have discussed some aspects of the construction industry: in particular, its structure, capacity and products. The demand and supply of its products and their importance to a national economy have also been highlighted. The importance of the industry is demonstrated by the way and manner national government influence the industry's activities in the process of a building.

In the UK, the central government's role as a major client of the industry is in decline as more private funds find their way into public sector construction projects. However, although there is a decline of the government's share in public construction investments, it still regulates the activities of the industry by its economic policies and legislation. The introduction of legislation on health and safety via the CDM is the latest government action in the latter direction.

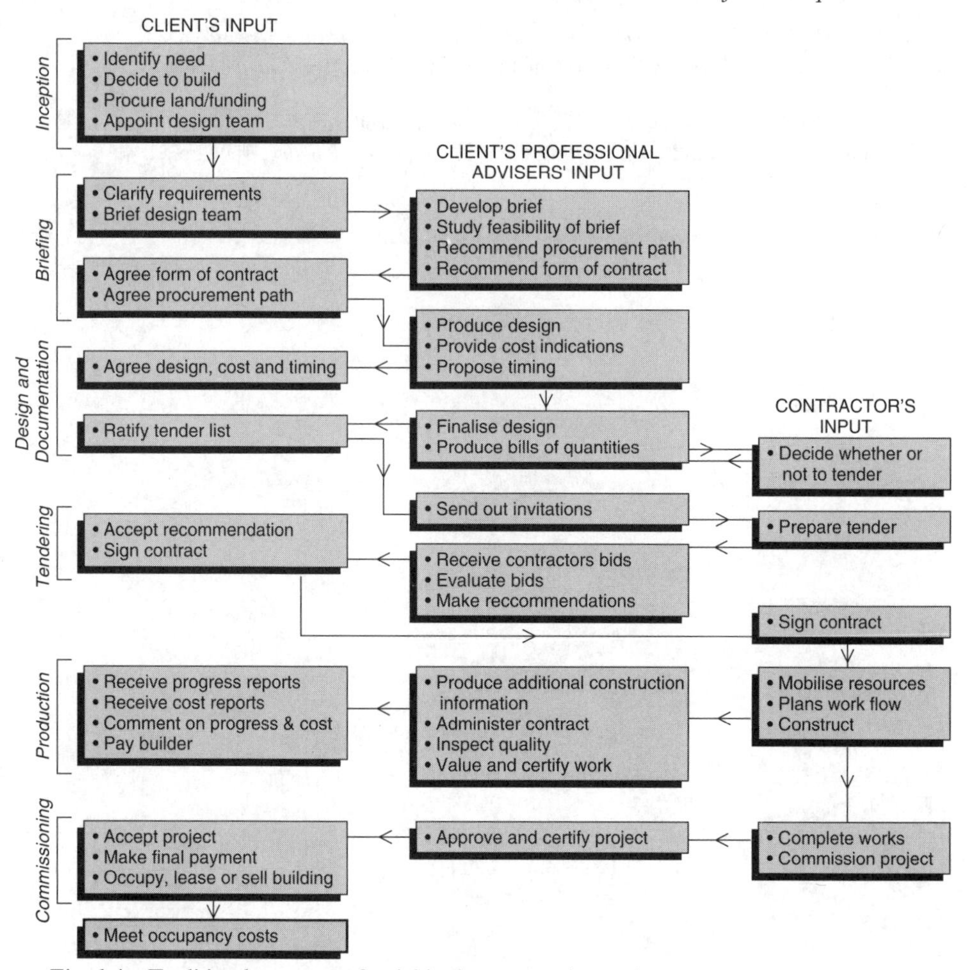

Fig. 1.4 Traditional sequence of activities in construction project

FURTHER READING

Bishop, D. 1974 Productivity in the construction industry. *Building*, August, pp. 77–8.

Bishop, D. 1994 Safety planning. *New Builder*, p. 121.

Bone, S.; Loring, J. 1994 How to be a planning supervisor. *The Architects Journal*, July, pp. 34–6.

Bone, S.; Loring, J. 1994 Meeting new responsibilities. *The Architects Journal*, July, pp. 23-5

Briggs, S. 1989 Managing health and safety in construction. *Chartered Quantity Surveyor*, pp. 24–5.

Hillebrandt, P.M. 1984 *Analysis of the British Construction Industry*. Macmillan.

Hillebrandt, P.M. 1978 Crises in construction. *Building Technology and Management*. March, pp. 4–6.

Hillebrandt, P.M. 1977 *Economic theory and the construction industry*. Macmillan.
Hillebrandt, P.M. 1974 The capacity of the building industry. *Building*, August, pp. 71–3.
Joyce, R. 1995 *The CDM regulations explained*. Thomas Telford.
Murphy, N.; Chapman, S. *et al.* 1980 Image of the industry. *Building*, May, pp. 31–4.

CHAPTER 2

THE ROLE OF PROJECT PARTICIPANTS

2.1 INTRODUCTION

A construction project is a task undertaken in the production of construction products. The term *project* in this context is being used for the total activity from inception to commissioning and occupation, involving an agreed and planned objective and total input of specialist participants and their interrelationships. It is a temporary non-recurrent activity which is started, implemented, evaluated and terminated. This activity is undertaken in response to a demand (direct, derived, individual or collective) for construction activity. Moreover, the activity is complex and, hence, necessitates the input of large numbers of participants with different disciplines to carry out the separate but interrelated functions of design, engineering, costing, pricing and production. The participants who are engaged to work on the project are mainly unaccustomed to working with each other and, hence, project activity imposes a special demand on team building and motivation. In addition, every participant should be made aware of all the governing conditions, objectives, responsibilities, relationship and other basic parameters of the construction project.

Construction projects vary considerably in size and complexity. Moreover, generally complex projects tend to be larger with large amount of services element. This complexity poses major problems of bounded rationality, risks and uncertainty. Hence, the project participants need to organise themselves properly and identify their respective roles, the project risks and uncertainties. They should also recognise the need to establish effective lines of communication which facilitates free flow of information throughout the duration of the project.

2.2 PROJECT PARTICIPANTS

The participants to construction project procurement are the client (who is the initiator), the multi-disciplinary construction consultants (who act as the client's professional advisers) and the building contractor (who constructs the building).

Together, this group of participants take on and manage the sequence of distinct but interrelated activities of the construction process from beginning to the end. A construction project, to all intents and purposes, is the production of capital goods and, like any other capital investment, involves careful planning and decision making. Hence, its production is the result of many months of process that required briefing, planning, engineering, designing, funding, budgeting and negotiating. Conversely, unlike any other capital investment, wrong decisions made earlier in the construction process are expensive to change and therefore a careful definition of the client's objectives and the responsibility of each participant at the pre-contract stage is required. Additionally, it demands a careful participative planning and thinking ahead to achieve some set goal against cost limit and time scale.

Construction projects generally are complex and composed of many activities. It is this complexity that calls for the input proposals of designers, contractors, suppliers and statutory authorities for their production. Although the procurement method adopted may vary the relationship of the participants, there will always be a proposer (client), designer (architect/engineer), construction team (builder), statutory authorities (gas, electricity, fire and water) and area local authority, as shown in Figure 2.1.

2.2.1 The client

The client is the key to the whole construction production process from inception to completion and at times to post-occupancy maintenance. Without the client there would be no construction project. Construction industry clients either identify user potential or create the need for the facilities and raise the necessary financial resources for their creation. They initiate the construction process by commissioning various construction professionals to build to specific requirements. While clients also select their projects timing, priorities, cost limit, and often determine the contractual methods, they do vary. Some clients are well informed and, hence, know what they want and take decisive steps to achieve it. Some know nothing about construction and need help and guidance to formulate their wishes and match them to the available budget.

Generally, when a client perceives the need for new construction or refurbishment, the decision to undertake it is made in the midst of environmental forces (political, social, technological, economical, educational, legal) and within a time scale (see Figure 2.2). As the initiator of the construction process, the client first develops the construction idea then studies the market to identify user or potential user demand (where the development is for sale, leasing or letting). Once a favourable demand has been established, the client carries out or gets others (construction professionals) to carry out feasibility studies, research for alternative courses of action and the development of brief for the design of the construction product. During the design and production phases, the client directly or indirectly monitors progress, time, cost, quality objectives and sanctions any necessary major variations to the design. Finally, on completion, it is the client who either disposes

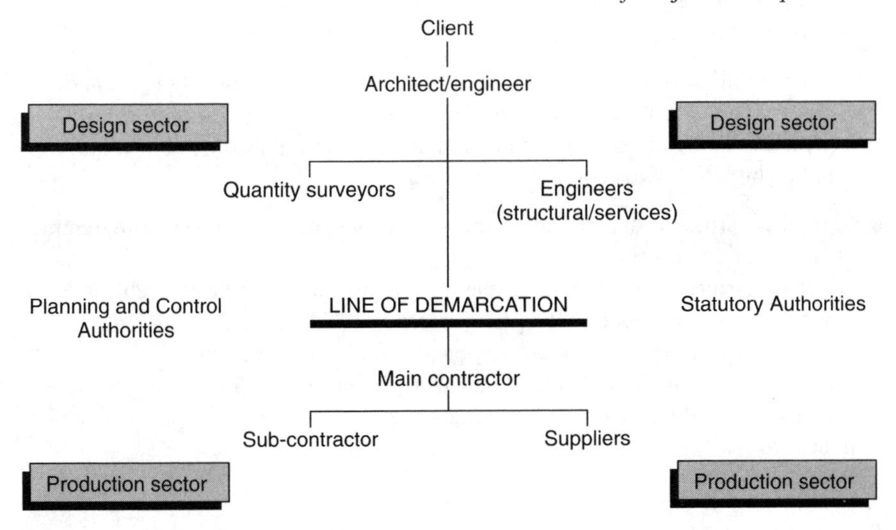

Fig. 2.1 Structure of project participants

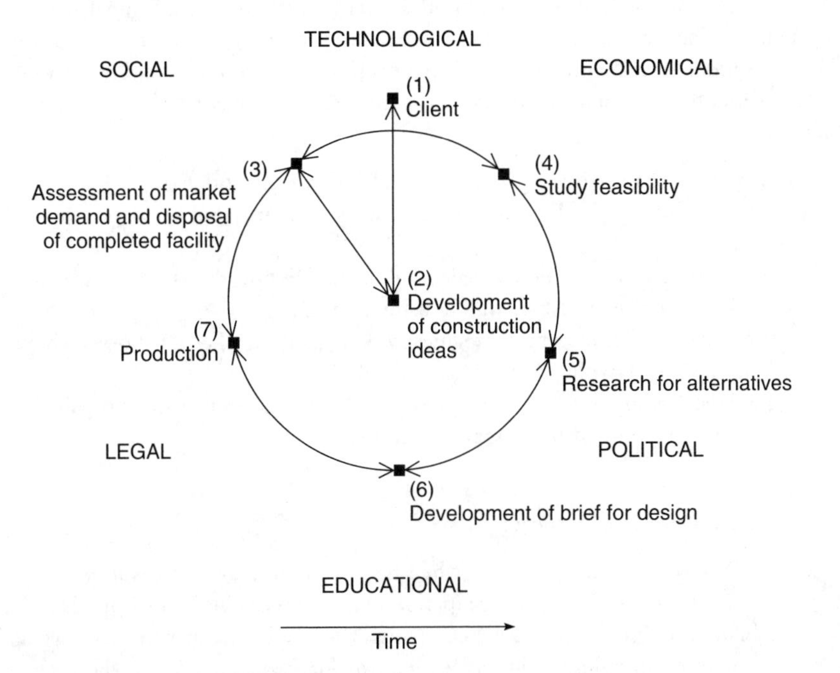

Fig. 2.2 Client's development cycle

of the product at the marketplace or takes occupation and bears the repairs and maintenance cost of his/her investment. Therefore, the construction industry looks to many clients for work and, generally, these may be classified as public sector clients or clients from the private sector.

Public sector client

These are public authorities whose operations are governed generally by Acts of Parliament. They act as agents for the central government who exercise control over their capital building programmes and expenditure. The key pubic sector clients for the UK construction industry are:

- Central government departments, who are responsible for their own programme of construction on projects.
- Local authorities, who are responsible for the provision of housing, schools, libraries, swimming pools, halls, sports centres and the like.
- Some health authorities, who are responsible for hospital buildings.
- Public corporations (e.g. British Rail and Air Transport boards), who are responsible for the provision of buildings and other construction products for their own use.

Private sector client

These are private companies which build for leasing, renting, sale or own occupation. The central government only exercises a limited amount of control over their operations (e.g. planning controls for proposed developments). The private sector clients for the construction industry are many and may be classified as follows:

- Multi-national companies (e.g. Ford, Cadbury's, ICI and Esso) who construct factories, production plants, offices and distribution depots for their own use.
- National companies (e.g. Tesco, Sainsbury's and Woolworth's) who construct buildings for their own use in warehousing and retail.
- Local property development companies, who construct offices, factories, shops and houses speculatively for hire, lease or sale.
- Private clients who construct new buildings, or extend, refurbish or repair the existing buildings for own occupation, letting, leasing or sale.

2.2.2 Client's role in construction

The client's development aim is met when he/she commits resources into construction production for own occupation (e.g. where an individual builds a house); for investment (e.g. where a company or an individual constructs a block of offices for own occupation, letting or leasing); or for speculation (e.g. where houses are built for sale). As the initiator of construction projects, the client plays a key role in the construction process. However, as all development plans involve a heavy expenditure of money, the economies of this plan require evaluation before the client becomes committed to substantial outlays. It is prudent, therefore, that the client pays particular regard to the following factors when drawing up a development plan.

Demand for the Product

As the objective of all construction development is some return (financial and non-financial), the client will not usually undertake any development without some prospect of this being present, and adequate. Therefore, while investment in a construction project may be a desirable proposition, before being over-committed, it is essential that the client considers *demand.* This demand should be economic demand (i.e. effective demand) which when considered in the context of the client's construction project, may be defined as construction product for which consumers are willing and able to pay. The effective demand, in turn, depends upon many variable factors and hence it is difficult to forecast. In addition, the problem differs according to whether the demand is individual or collective; or whether the client is building speculatively for unknown consumers or by contract for specific consumers. But since consumer demand regulates the execution of construction projects, it is important that great care is exercised by clients when forecasting the future levels of demand for their planned construction programme.
Furthermore, demand is a dynamic concept in that it is constantly shifting in response to changes in the economy at large. But as the consumer demand for any individual or family depends on income distribution and spending habits, the most important factors taken into account in any demand forecast are changes in interest rates, credit and mortgage availability; and at a more local level, migration, employment and population changes. What is more, by applying professional knowledge and analysing critically the answers to the following questions, the client will be able to tackle the demand forecasting process with some confidence of success.

- Will there be more/less household formation particularly by young couples?
- Will the proportion of new households choosing and being able to afford owner occupation change?
- Will the relationship between local house prices and incomes be more/less favourable?
- Will access to other tenures (e.g. renting/leasing) become easier/more difficult?
- Will the local employment situation change, and what effects will this have on migration to and from the area?
- Will there be more/less potential in-migrants from other areas as a result of, for example, changing local economy?
- Will there be more/less housing opportunities in the existing stock as a result of owner occupiers moving out, refurbishment or clearances acquisitions?
- Will there be more/less sales by private landlords?
- Will there be more/less new buildings or cheap/attractive housing opportunities either locally or in neighbouring districts?

To a certain extent, past trends can be extrapolated into the future, thus helping clients to find quick answers to some of the above questions. However, the above approaches can give inacurate predictions due to the ever-changing situations in the economy, human life and so on. Generally, demand for construction activity is governed by the level of economic activity and varying economic and social policies pursued by governments. It is, therefore, also important that the client should be able to predict, as accurately as possible, any likely changes in government spending plans and taxation policies as these actions influence consumer spending power which, in turn, influence demand for construction products.

Availability and cost of land

Dependency on land is what makes construction development different from other investment activities. Usually, an efficient and effective construction programme depends upon the client having sufficient land available in the right place. If land is not already owned, it should be procured at the right time, place and at a realistic price. Also, the condition of the land and its suitability for the project should be studied to determine the extent of site preparation needed before the development. The purpose of this exercise is to work out the true cost of the land as costly site preparation may cause reasonably priced land to become very expensive in the long run. It is, therefore, essential that the following factors are considered before any land is purchased.

1. *Size or area of land:* Under the UK planning regulations, the size of building permitted on any particular plot of land is determined by the area of land. The size of the plot of land must therefore be sufficient to allow the intended development to take place.
2. *Location and accessibility:* As land is fixed, its location is of crucial significance to its use and value. A good land location signifies its quality and, hence, influences the profitability of its use. Accessibility, on the other hand, is the term used to describe the varying locational advantage or disadvantage of land. Where the location affords easy access to resources, customers, services and reduces movement and transportation costs, these factors lead to a profitable development. Furthermore, where access to the land during development involves construction of unusually long access roads or creates difficulty in the movement of plant and materials to the land, the cost of the construction project, as a result, increases.
3. *Price:* The price the client pays for any particular plot of land is determined by the interactions of supply and demand for land for development in a particular area. The size of the required plot of land also reflects its price.
4. *Required infrastructure:* The installation of long runs of pipework for gas, water and new sewers and the construction of a pumping station or an electricity sub-station can add a considerable cost to a construction project.
5. *Legal constraints:* Legal constraints such as restrictive covenants, easements, rights of way, right of light and so on can affect the viability of the intended construction project.
6. *Current and future development:* Current and possible future development in the neighbourhood which might create such things as noise, pollution, nuisance and contamination can blight the construction project and this needs to be taken into account when purchasing a plot of land in such a neighbourhood.

7. *Physical key features:* Physical key features such as ancient monuments and listed structures (which require retention) and services running over and underground (which needs removal, renewal or diversion) may add extra cost to the construction project.
8. *Physical characteristics of land:* Soil condition, formation, ground-bearing capacity, water table levels, flood hazard and so on may require the adoption of expensive precautionary measures. Additionally contaminated or filled land may be uneconomic to develop owing to the expensive investigative and preparatory work these may require.
9. *Site preparation:* Site preparation such as demolitions, decanting/rehousing of occupants and road closures can delay the commencement date of a construction project and possibly increase its overall cost.
10. *Planning permission:* A development land has not the same meaning to a prospective client and a local planning authority. Therefore, clients must ensure that they can procure planning permission for the intended development on the land before embarking on time-consuming, expensive and uncertain moves.

Availability of finance

No construction project can start without adequate funds to finance it. Where the client has not the funds for a development, he or she has to make funding arrangements to meet its cost. If a client intends to sell the finished product, he or she requires finance only for a limited period (i.e. short-term finance). On the other hand, if a client proposes to retain the completed facility as a permanent investment, the need to raise short-term and long-term finances is essential.

1. *Short-term finance:* This financial arrangement is made for the purchase of land and paying for the cost of construction including all professional fees. In effect, this is fund needed to construct the facility before any income is received. The duration of credit facility may last for three to five years or until the product is completed. The primary sources of short-term finance are internal sources (in the case of major concerns), the commercial and merchant banks (subject to government controls). These banks are prepared to lend on less risky schemes and to clients with sound track records. Short-term finance is relatively expensive as a client can offer only limited security (land and unfinished construction product). However, the rate of interest charged by the banks varies and depends largely on a client's financial status, quality of construction project for which finance is required and the development period.
2. *Long-term finance:* Clients may raise long-term finance either by selling outright an interest in a development or by borrowing against the security of a finished product. Long-term finance is secured on a completed construction project and, hence, capital security is much greater and as a result, attracts a lower rate of interest. The primary sources of long-term finance in the UK is normally the financial institutions such as insurance companies, pension funds and investment

trusts. Unlike the banks which seek liquidity and therefore are interested in lending short term, the financial institutions seek security and growth of income and therefore tend to lend on a long-term basis.

A client may also adopt the *sale and lease back* approach if he or she wishes to retain a leasehold interest of the completed product. Under this arrangement, the client sells the finished product but immediately leases it back at a rent representing a proportion of the full rental value, thereby allowing the client to retain an interest in the product and gain benefits from a share of the rental income.

3. *Partnership schemes arrangement:* Where a client has devised a development plan but lacks suitable land and/or the necessary funds for the development, he or she may enter into partnership with a landowner (e.g. area local authority) and financier (e.g. financial institution). Under this arrangement, the local authority provides a plot of land, an infrastructure and a development brief; the financial institution provides the necessary funds and the client (at times contributing to land acquisition) provides the expertise of managing the construction process. On completion of the project, the local authority receives ground rent plus a share in the equity of the development; the financial institution receives interest on funds advanced plus a share in the equity; and the client receives a share of the equity.

Appointment of professional adviser

As a construction process is complex, a client may require the input of a professional adviser in order to achieve his or her development aims. Construction professionals advise on, organise and participate in all stages of briefing, design and construction. Therefore, traditionally, clients procure advice from construction professionals (namely the architect, quantity surveyor, structural engineer and services engineer). Their separate functions contribute to the conceptual thinking, produce alternative sketch designs, help to formulate aims and requirements and assess the viability of the construction project. In all, construction professionals assist clients to obtain, within their budget, the high-quality project to which they aspire. When a client is developing to let, sell or lease, then the services of a valuer is also required to provide advice on the probable level of development value and letting or sale income.

Formulation of the project brief

In setting the construction scene, traditionally the architect is the first construction professional a client may wish to appoint. On appointment, the client briefs the architect on his/her requirements in terms of time, cost, function and quality objectives. The financial implications of a client's other requirements such as plant, standards of finish and so on should be set down at this stage. The clarity and adequacy of brief is a prerequisite for the success of every construction project because it facilitates its full development and enables the client to explore the many courses of action open to him or her during the planning stages of the project. With professional guidance, the client should be able to examine the alternatives and establish their economic viability and technical feasibility. If this initial exercise proves that the construction project is a practical and economic proposition, then the client should give the scheme 'the go ahead' and, thereafter, the project proceeds through successive stages towards completion.

The need for quality assurance

As organisations face new demands and expectations in today's changing competitive environment, there is a need for integrating total quality in service/ product provision. This is a strategic move which addresses the marketplace and enables products and services to satisfy fully the demands of (internal and external) customers. Within the construction process, there is no contravention of quality standards until the site is opened and construction starts. Hence, traditionally, a client controls quality by the standards he or she specifies and also through inspection by his or her professional advisers and, perhaps, by the site supervisory or inspection staff appointed to oversee to the achievement of those specified standards. Therefore, the achievement of the standard required by the client depends on both the contractor and the professional advisers, and for this reason it is in the client's interest to request quality assurance statements from his or her professional advisers as well as from the building contractor.

The production of such a statement is an indication that an establishment is conscious of quality and adopts a quality assurance practice that promotes consistency of performance. It may also indicate the existence of a management system capable of applying appropriate management checks throughout all work stages, be it a design service, a product-manufacturing process or a construction process and correcting deficiencies as and when they occur. In addition, this gives the client a guarantee that the construction project design and specification are of high quality and that the project will be produced under satisfactory conditions of quality control to satisfy the following criteria:

- Construction project properly constructed and fit for a defined purpose.
- Ability of construction project to continue to perform as required for a reasonable period of time.
- Conformity of a construction project to a specified standard of amenity, comfort and aesthetics.

It must be emphasised, however, that there is a cost associated with maintaining high quality but quality control is essential and, hence, necessitates a built-in quality allowance within a client's budget.

2.2.3 Client's needs and expectation

After the client has appointed all the professional advisers, they are expected to use their professional skills, experience and judgement in the following:

Selection of procurement method

In principle, the most suitable procurement method adopted for a client's construction project depends on:

- Complexity and scale of the project.
- Expectation of specific performance requirements.
- Level of risk client is willing to accept.
- Necessity for competition on price and time.
- Necessity for accountability on the part of those concerned in its administration.
- Pre-commitments and existing relationships.
- Personal perceptions by management.

Translation of brief

It may not always be possible for an inexperienced client to formulate the project's brief adequately and, being aware of this deficiency, he or she expects the professional advisers to be able to provide a technical interpretation of his or her requirements. The client therefore, expects a good-quality design in terms of aesthetic merit, functional and durability from the original concept.

Reliable initial cost advice

The client would like to be given assurance that the final cost of the construction project will be very close indeed to the original estimate. Such an assurance will enable the client to know the financial commitment of the project prior to commencement on site.

Definite date for occupation/disposal

The client needs assurance that the construction project will be commenced and completed on a set date. This assurance will help him or her plan when use can be made of the facility or what income (sale or rental) to expect.

Low cost maintenance product

Where a client opts for low maintenance cost, it is an indication of his or her willingness to incur a high initial construction cost. Therefore, the design and material/component selection should reflect this desire.

Suitable contractual method

Depending on the circumstances (e.g. whether time or strict financial control is of the essence), a client expects the professional advisers to select a method of contract that suits the project strategy.

Adequate contract documentation

Inadequacy of contract documentation is one of the prime causes of construction contract disputes. Generally, the resolution of construction disputes can be time consuming and costly. For this reason, the client expects that the professional advisers will avoid all forms of construction disputes by production of adequate contract documentation for the construction project. The client also expects that complex contractual matters such as insurance, bonds, collateral warranties and damages for late completion are well covered in the contract documentation.

Selection of a suitable contractor

A good contractor for a construction project is an important factor contributing to its successful and timely completion. Therefore, a client would normally expect professional advisers to select a reputable building contractor with a good financial standing and potential resources to complete the project successfully on time, to the client's satisfaction and without expensive contractual claims.

Value for money

The client expects that design decisions and material/components selections by the professional advisers will give value for money. This value for money should not be in terms of cheap solutions but rather well-calculated solutions that will produce a project that will stand the test of time, give user satisfaction and benefit the community for the cost and efforts incurred.

2.2.4 Client's responsibilities

The foregoing client's expectations put a lot of pressure on the professional advisers to seek feasible and economic decisions for the client's construction project. For example, a good choice of timing of scheme design, contract documentation and site production may affect appreciably the cost of interest payment on the project finance. Furthermore, design for buildability may also facilitate construction, and reduce material wastage and production time. Additionally, while the temporary and dynamic nature of the construction process create problems of motivation and control, construction professionals must endeavour to overcome this problem and unite in their efforts to find satisfactory solutions to the problems posed by the client's construction project. Nevertheless, while a client can be tough, demanding and expect better value for money, he or she would not achieve this without a price of cooperation. The client should therefore recognise that the efficient performance of construction professionals depends partly on the client's coöperation and partly on the importance that is attached to the following duties and responsibilities.

Provision of adequate brief

The client must be able to collect and analyse all relevant information and provide an adequate statement of requirements that is capable of formulation and development. Inadequate briefing complicates design, causes abortive work, poor documentation, excessive variations, delays, additional cost and poor quality construction product.

Clarity of brief

In order to reduce or eliminate a last-minute change of mind, the client must be clear and precise in the brief. The client must also try to limit any variations to the agreed design to a minimum in order to avoid abortive expensive work and consequential delays to the contract.

Supply of prompt decisions

The client should be able to take quick decisions on various matters submitted to him or her at design and production phases of the project. Such prompt responses avoid hold-ups and maintain momentum of design and production.

Adequacy of finance

The client should consider the financial aspects of the proposed construction project and ensure that adequate funds are available to meet progress and professional services payments. Funds should also be available to honour extra contractual claims and professional fees associated with it.

Disruptions

The client's decisions and/or actions should not cause any disruption to the progress of the works. Where the client intends to carry out any section of the works personally, this should be properly planned well in advance of the start of the project to ensure its smooth incorporation without disruption to the contractor's planned production.

Avoidance of interruptions

The client should leave the professional advisers to make technical decisions which are within their expertise. Frequent interruptions and challenge of such technical decisions by reason of high cost does not encourage the provision of sound technical advice. In addition, the client should not concentrate on cost-cutting actions, particularly in design, which might lead to lack of attention to key activities and good basic detailing.

Land Acquisition

The client must consider his or her legal titles or responsibilities regarding the land for the construction project and, where necessary, must engage the services of a legal expert.

Insurance

The client must arrange, or instruct others to arrange on his or her behalf, any necessary insurances for the construction project.

2.2.5 The design team

The architect

Traditionally, the design function in the construction process is the responsibility of an architect who is a professionally qualified person whose role is to interpret the client's project requirements into a specific design or scheme. Design is taken to include appearance composition, proportion, structure, function and economy of product, but in addition the architect performs the function of obtaining planning permission for the scheme. In most times, too, the architect supervises and organises the entire construction process, starting with consultation with the client and ending with commissioning. As an established practice, the architect plays the leading role in the construction process. He or she collects, coordinates, controls and disseminates project information to all project participants. As a project team leader, the architect performs various functions in all the stages of the construction process, which includes:

- Ascertaining, interpreting and formulating the client's requirements into an understandable project brief.
- Designing a building to meet the client's requirement and constraints imposed by such factors as statutory obligations, technical feasibility, environmental standards, site conditions and cost.
- Bringing together a team of construction professionals such as the quantity surveyor, structural engineer and services engineer to give expert guidance on specific points of the client's construction project.
- Assessing client's cost limit and time scale, and specifying the type and grade of materials/components for use on the construction project.
- Preparing production information for pricing and construction and inviting tenders from building contractors.
- Supervising the production on site, constantly keeping client informed of the project's progress and issuing production instructions as and when required.
- Keeping the client informed of the status of project's cost and advising on when payments should be made or withheld.
- Advising on the conduct of the project generally and resolving all contractual disputes between client and the building contractor.

- Issuing the certificate of completion, the certificate of making good defects and the final certificate for payment.

Generally, the architect acts as an agent for all purposes relating to designing, obtaining tenders for and superintending the construction work for which he or she has been commissioned. To be able to perform the above function efficiently, the architect must possess, among other things, the attributes of foresight, an understanding of construction materials, communicating and coordinating abilities, essential design skills and an ability to design within a set budget.

The architect may be an individual in private practice either as a sole practitioner or as a partnership and adheres to a very strict code of professional conduct. Being under a strict code of conduct and, moreover, forbidden to advertise his or her services, the practising architect may obtain commissions by the following means:

- By competition (i.e. taking part and winning design competitions initiated by clients).
- Through personal contacts with friends, relatives and acquaintances and social contacts with clubs and associations.
- Recommendations by satisfied clients.
- By invitation through reputation gained and professional status in the business community.
- By continuous employment as a consultant for a client with large programme of development projects.

Normally, once the decision to build has been made, the choice of an architect is the employer's first priority. However, when appointed, the architect does not perform the design and supervisory functions of design alone; rather the architect recommends the appointment of other professionals, namely the quantity surveyor, the structural and services engineers and clerk of works to collaborate with him or her in the performance of those roles. Together, these professionals apply their skills in their respective fields and assist the architect in solving the problems posed by the client's construction project in terms of function, stability and cost.

The quantity surveyor

The quantity surveyor is responsible for the study of the economies and financial implications of a construction project and, hence, he or she would be the appropriate construction professional to advise client/architect on matters relating to the economies and cost of a proposed construction project. Traditionally, quantity surveyors organise themselves into small practices; however, many are now to be found in contracting and client organisations. Those in private practice are mostly chosen and appointed by clients on the recommendation of an architect.

As cost is one of the deciding factors in most construction projects, the quantity surveyor is brought in at the earliest opportunity to advise the client or architect on the cost of various schemes proposed. The quantity surveyor is also able to perform several functions on construction projects, and these may be summarised as follows:

- Preparation of preliminary cost advice and approximate estimating.
- Preparation of cost plan and carrying out cost studies (investment appraisal, life cycle costing and the like).
- Preparation of contract documentation for contractor selection and construction project administration.
- Evaluation of contractors tenders received with recommendations for acceptance or rejection.
- Preparation of cash flow forecasts and institution of post-contract cost monitoring/reporting mechanisms.
- Valuation of variations that arise as the works proceed and preparation of interim valuations at regular intervals.
- Preparation of periodic cost report for the architect or client.
- Preparation and agreement of final account with the contractor.
- Evaluation and settlement of contractor's claim for direct loss and/or expenses.
- Settlement of contractual disputes.

The structural engineer

The structural engineer acts as an adviser to the architect on all structural problems such as stability of the structure, suitability of materials proposed, structural feasibility of the proposed design and sizes of structural members for a construction project. Normally, the structural engineer submits his or her various structural calculations to the area local authority for approval at the same time as the architect submits his or her drawings for building regulations approval. In addition, the structural engineer performs structural design and supervises his or her specialist area of the construction project during production on site.

The services engineers

Like the structural engineer, the services engineers (plumbing, electrical, heating and ventilating, air conditioning, sanitation, lifts and escalators and so on) contribute to the building design process to ensure that thermal and visual comfort are achieved effectively. For this reason, they analyse the client's requirement and priorities and advise the architect on the most appropriate design solution. They prepare diagrams of their proposals or services layout of the proposed construction project on separate drawings and the architect includes these in the tender drawings sent out to contractors for competitive bidding. Once the services engineers have made their contributions to the design, they ensure that their contributions have been correctly interpreted, installed and commissioned. Where services engineers' design layout causes any structural problems, the advice of the structural engineer is sought. There is also a need for the architect to coordinate the route of pipes, cables and ducts for various services on the project.

The duties and responsibilities of the structural and services engineers include the following:

- Providing specialist advice and assisting in the design of the construction project within the scope of their respective specialist fields.
- Producing calculations or other relevant data to assist in the design, cost planning, and the assessment of suitability of materials/components and the like.
- Supervising their respective specialist fields of the project and modifying or redesigning work whenever required.

2.2.6 The role of construction professionals generally

Together, the architect, the quantity surveyor and the structural and services engineers are known as the 'construction professionals' or 'the design team' who offer design and cost advice services to the client for a fee related to the size and complexity of the project, the terms of appointment and the services required. Usually, the degree of their relationship with a client depends on:

- Knowledge of the client on construction matters and his or her ability to influence projects to suit the strategy.
- Size and structure of the client's organisation and the availability of experienced personnel to coordinate the proposed construction project.
- The complexity and scale of the construction project. Where the project is complex, constant contact with the client will be essential to ensure incorporation of all clients requirements into the design.

The need for this professional role on construction projects has arisen on various grounds. One is that clients normally require advice because they do not have the expertise to develop the brief, produce a sound design and supervise the construction. A second major reason is that the design should take matters such as environmental issues, the needs of society, functional requirement of the client and safety into account, and only construction professionals are able to consider these matters adequately. For this reason, the construction professionals perform an important advisory role: they organise the construction process and participate in briefing, design and construction. In the process, they consider all matters that affect the quality of the environment from the point of view of the ultimate user and of society as a whole. They also ensure that their design solution gives the client value for money and affords the most economic production process.

2.2.7 The builder

The production aspects of construction projects are undertaken by building contractors who are essentially commercial companies that contract to construct development projects. Although many major contracting establishments are able to undertake both design and production work, their primary function is to build and to organise their considerable resources basically as a manufacturing organisation. Therefore, unless building speculatively, building contractors do not initiate construction process but only build construction projects for clients. While in the business of building for construction clients, the building contractor obtains work by the following means:

- Recommendations from past and/or satisfied clients.
- Successful inclusion on client's tender list on application and after vetting.
- Personal contacts with people in high places (e.g. politicians, directors in private and public sector organisations).

Generally, their objective is success in business, which should mean a satisfied client and potential repeat business, satisfied client professional advisers, enhanced reputation and satisfied shareholders. Construction companies are many, of various sizes and in a wide geographical area. A firm may choose to operate in a particular locality or region, but some operate nationally and internationally. The main building contractor takes responsibility for entire projects but does not execute every aspect of it; rather, it sublets parts of the works to the many sub-contractors who are specialist firms of tradespeople and suppliers who manufacture and install the constituent parts of construction product.

In the UK, companies offering sub-contracting services in the construction market may be categorised into two groups, namely, the group offering labour only service (i.e. labour only sub-contractors) and the group offering labour and material. The latter group may be further classified either as *domestic sub-contractors* (i.e. organisations that main contractors engage to execute sections of the project) or *nominated sub-contractors* (i.e. organisations nominated by the client to execute a specific element of the project). These specialist sub-contractors serve the needs of a large number of main building contractors. They thrive because of their known reliability in performing some specialised part of the work or installations at a lower cost than the main building contractors carrying them out in-house.

Duties and responsibilities of the contractor commences upon invitation to tender and includes the following:

- Carrying out a full site investigation prior to submission of tender to ensure that the bid includes all the cost of contractual risks and problems.
- Submitting a priced bills of quantities for examination and/or correction of any errors when required by the architect.
- Planning and programming the works and reprogramming thereafter whenever unforeseen events frustrate the programme.
- Controlling directly employed operatives, sub-contractors, suppliers, materials and plant for the execution of the project to programme and cost.
- Coordinating the efforts of all operatives and ensuring that the completed works comply with the contract specification and are also to the satisfaction of the architect.
- Notifying the architect of information requirements, delays to the construction programme, discrepancy between contract documents, direct loss and/or expense sustained and so on.

- Paying the wages of directly employed operatives, sub-contractors and suppliers in time to avoid conflicts over payment.
- Supplying all the information required by the client's professional advisers for the proper administration of the works.
- Taking steps to carry out the contractor's obligations to rectify all defects on completion of the works.
- Providing post-occupancy repair and maintenance service if so required by the client.

2.2.8 Public sector agencies

Public sector agencies are organisations that have been set up with a specific authority either to run public utilities or to provide local services. Some of these agencies participate in construction projects and, hence, need to be consulted for guidance. Public sector agencies that contribute to the success of construction projects are:

1. *Statutory authorities:* The statutory authorities that participate in construction projects are water board (for water supply), gas board (for gas supply) and electricity board (for electricity supply). These bodies take part of the construction process at both design and production phases and offer useful technical advice in their respective fields of expertise. For this reason, it is always beneficial for the client to give them early information of any proposed development plans and work closely with them thereafter for their guidance.
2. *Local authorities:* In the UK, an area local authority has control over the construction of buildings by powers derived from Acts of Parliament and hence, it is the only body that grants planning permission for the execution of construction projects in its area of jurisdiction. For this reason, application forms and drawings must be submitted to the area local authority and an approval must be obtained before commencement of site production. Designers have to satisfy the planning authorities that the new building will fit in with its surroundings as well as meeting other stipulations under planning control such as height, location, access and density. At the same time, the area local authority has the power to inspect construction projects as they progress at specific stages on site to ensure that a particular design solution meets the required standards of construction and performance.

   Generally, approvals and controls do not prevent failures in buildings and the fact that a particular design has passed the planning approval is not an indication that the materials and details specified will perform as intended. Therefore, when failure occurs through bad design, the blame rests squarely with the designer and not with the local authority or its officers (namely, the building control officers). The exercise of strict building control is essential for the following reasons:

   - Keeping available for development any sites that will be required for specific purposes.

- Ensuring that an available plot of land is put to an intended use for the benefit of the community at large.
- Ensuring that buildings and/or industries are located in areas earmarked for specific developments to ensure coordinated development and/or complementarity of industries.
- Health and safety of occupants ensured through adoption of improved design and construction practices.
- Approvals and controls raise standards of development and ensure the safety, health and welfare of site operatives/ultimate consumer.
- Preservation and protection of assets; and creation of amenity.

Like the statutory authorities, local authorities influence design and production functions of almost all construction projects in their area of administration.

3. *Fire authority:* The fire authority provides advice on fire protection of occupants, contents and structure of a building. Usually the principal source of fires in buildings is their contents. However, the combustibility of the content of buildings varies considerably; therefore, when consulted the fire authority will provide information on structure suitable for:

- Controlling fire within a building.
- Providing safety for occupiers of building in case of fire.
- Providing protection for property and contents within reasons.

SUMMARY

In this chapter we have discussed the role of all the participants of a construction process. The client is certainly the key player in all construction projects and the objective and rewards that influence his or her decision to invest in construction products should be clear before the decision to build is taken.

Since the procurement of construction is risky and requires heavy expenditure, measures are taken by clients to reduce risks and thereby make their investment in construction successful. Generally, the client's professional advisers assist in the construction production and management; however, as their role is of an advisory nature, the financial success of all projects rests on the client's own research and intuition in making the right decisions and on the part the client plays in getting the project built.

FURTHER READING

Balchin, P.N.; Kieve, J.L. 1979 *Urban land economics*. Macmillan.

Ball, M. 1988 *Rebuilding construction – economic change in the British construction industry*. Routledge.

Forster, G. 1978 *Building organisation and procedures*. Longman.

Hillebrandt, P.M. 1984 *Analysis of the British construction industry*. The Macmillan Press.

project strategy through which potential risk may be uncovered and possible responses framed.

The range of risks normally identified in construction projects include the following:

1. *Fundamental physical risks:* Fundamental physical risks include an outbreak of war, hostilities, nuclear accidents, damage by floods, storm and fire, exceptionally adverse weather and industrial action.
2. *Legal risks:* Legal risks consist of injury to persons and damage to buildings due to subsidence, vibration and similar events during production.
3. *Construction-related risks:* Construction-related risks include shortage of resources (labour, plant and materials), late completion, defective design, delayed possession of site, interference by numerous variation in quantity of work, delay in the issue of instructions, discrepancies found in contract documentation, postponement of site activities and effects on contractor's production by artisans engaged directly by the client.
4. *Price determination risks:* Price determination risks consist of errors in estimating, erroneous adjudication, inaccurate assessment of project risks and incorrect forecasting fluctuations on cost of resources.
5. *Contractual risks:* Contractual risks include uncontrolled delays, late payment, poor performance of project participants, faulty workmanship by untried labour, contractual claims and disputes and overrunning on project programme
6. *Performance risks:* These may be the result of productivity of labour affected by low morale, strikes, labour disputes, inadequate production planning, inadequate safety measures, production accidents, management inefficiency and operations that prove to be more difficult than expected.
7. *Economic risks:* Economic risks consist of inflationary pressures, rising cost of resources, high interest rates, funding delays and budget overruns.
8. *Political risks:* Political risks consist of environmental issues and organised protests, general public disorder, changes in government and changes in taxation.
9. *Commercial risks:* Commercial risks may be caused by market recession, contracted demand for type of development, strong competition from rivals (e.g. for resources and market share) and rivals under-cutting the price of finished products.

3.3 CONSTRUCTION PROJECT RISK ANALYSIS AND MANAGEMENT

By its very nature, there is no construction project that can be undertaken without an element of risk. Therefore, there is a need to identify the inherent risks involved in a specific project and manage them accordingly. Construction Project Risk Analysis and Management is a technique which considers the whole activities necessary for the identification, assessment, analysis and management of the risks economically. It is an integral part of good project management which should be used on all major construction projects and can provide proven benefits. However, it must be remembered that the aim of risk management is not necessarily the elimination of the risk factor. Rather it is a systematic way of looking at the area of risk, containing it to a level or minimising the risk to make it commensurate with its significance and the cost of control. To do this one needs to:

- Be aware of the risk.
- Identify the risk factor.
- Analyse the risk.
- Formulate appropriate management responses.

3.3.1 Risk awareness

The primary objective of risk management is to raise awareness of the areas of risks to which a project participant is exposed in the execution of the project. The project participants' awareness that risks are almost always present and, hence, are inherent in everything they do and every decision they make motivates them to identify and analyse the risk factors prior to any action. Project participants' awareness of, for example, fire, flood, explosion, subsidence or accidents enables them to take the appropriate measures to mitigate these risks. For this reason, the implementation of any effective risk management in construction rests on the awareness of risk factors inherent in construction projects.

3.3.2 Risk identification

The identification of the risk factors is an important and difficult task but must be undertaken before the institution of any risk-controlling or allocation mechanism. Risk identification normally involves detailed examination of the project strategy, project participants' perception of this strategy and their ability to perform to uncover the risk factors. This, therefore, means that the process of risk identification should take place at the earliest part of the project's life (ideally, prior to the feasibility appraisal of the project) and may be achieved by:

- Interviewing key project participants.
- Brainstorming meetings with all interested parties.
- Reviewing past corporate experience (if appraisal records are kept).
- Reliance on past experience of the risk analyst.

Having identified the risks, the next stage is the initial assessment that categorises the risks into high/low, probability/frequency of occurrence and major/minor impact on the project should the risks eventuate.

3.3.3 Risk analysis

The aim of risk analysis is to quantify the effects of the risk factor identified on the construction project. The quantification is essential as it enables the magnitude of the risk factor to be known. Moreover, this revelation assists the project participants to decide whether to take on the risk or avoid it through the combination of professional judgement and statistical methods. However, the choice of technique will usually be constrained by the type of risk, past experience and expertise, impact of the risk factor and occurrence probability.

The main techniques currently in use are:

1. *Sensitivity analysis:* This method seeks to determine the effect on the whole construction project of changing one of its risk variables while keeping all other variables constant. In its application, each risk – such as delay, inflationary pressures, high interest rates, cost of materials and so on – is considered individually and independently with no attempt to quantify probability of occurrence. If the project risks are tested in that way, the project participants are able to find how sensitive it is to change one risk variable (i.e. how the effect of a simple change in one risk variable can produce a marked difference in the project outcome). In the process, the technique enables the identification of the risk factor(s) whose variations or combined variations have a potentially severe impact on the cost, timescale or economic return of the project. In practice, a sensitivity analysis is performed for a large number of risks to enable project participants to identify high sensitive risk factors on which efforts should be concentrated.
2. *Probability analysis:* This type of analysis comprises the development of a risk analysis model to determine the probability of risk eventuating. The most common form of probability analysis adopts *Sampling techniques* and is known as *Monte Carlo Simulation*. In brief, the Monte Carlo Simulation makes the assumption that parameters subject to risk can be described by probability distributions. Hence, it relies on the random calculation of values that fall within a specified probability distribution often described by using three estimates: minimum or optimistic, mean or most likely and maximum or pessimistic. The mean of the distribution is calculated from the following formula:

$$\frac{O + 4M + P}{6}$$

where O is the optimistic point prediction, P is the pessimistic point prediction and M is the most likely point prediction.

By assessing the range of values for the risk factors under consideration (together with the probability distribution most suited to each risk) and selecting a random value from the probability distribution for each risk within its specified range (using a random number generator), the overall outcome of a construction project can be devised using the combination of values selected from each of the risks.

3.3.4 Risk management

The risk management process involves the formulation of management responses to the risks, using the information collected during the risk analysis phase described above. This is done by reducing the risk where it is advantageous to do so, or monitoring and managing those risks which remain. Risk management may start during the analysis stage as the need to respond to the risks may be urgent and the solutions fairly obvious. Once started, it remains a continuous process throughout the project's life cycle. In the risk management process, management's response to risks may be defined either as *immediate* or *contingency*. These are defined as follows:

- *Immediate response* is an alteration to a project plan such that the identified risk is mitigated or eliminated.
- *Contingency response* is a provision in a project plan for a course of action that will only be implemented should the adverse consequences of the identified risk eventuate.

Range of Options

Generally, the range of options open to one exposed to risk are as follows:

- Acceptance of the risk.
- Transfer risk to insurers.
- Allocate the risk to a third party.
- Reduction of the risk.
- Removal of the risk.

1. *Acceptance of risk:* This is where the client, for example, has decided to take on the risk factor(s). This response may flow form:

   (a) Ability to manage risk to advantage and acceptance also of the inherent liabilities.
   (b) Inability to allocate risk to a third party.
   (c) Inability to avoid risk factor.

   In this option, the benefits that can be gained from accepting the risk should be balanced against the liabilities.
2. *Risk transfer:* Risk transfer is a form of risk handling which involves shifting the risk burden from one party to another. In construction projects, this may be accomplished through contract conditions or by insurance. The participants to whom the risk burden has been transferred generally respond by including an appropriate allowance in the cost estimate to cover this risk. The extent of this allowance generally reflects on the magnitude of the risk (i.e. the greater the magnitude of risk, the greater the amount of risk allowance and vice versa). The most common risk transfer arrangements (see Figure 3.1) are:

   (a) Client to designer.
   (b) Client to contractor.

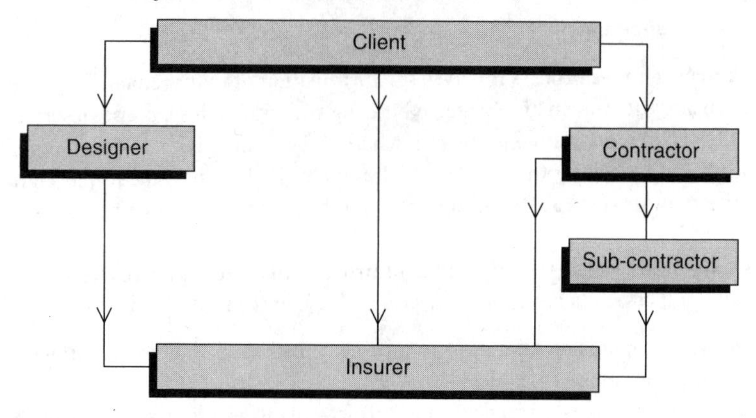

Fig. 3.1 Transfer of construction risks

(c) Contractor to sub-contractor.
(d) Client, designer, contractor or sub-contractor to an insurer.

The purpose of insurances is to convert the risk into fixed cost. In this manner, the real cost of the risk is known. However, the following information should be provided for any risk transfer arrangement in order to comply with general insurance laws and practices.

(a) Assurance that the risk is identified, sufficiently described to facilitate pricing and provide a framework in which an adequate price for the assumption of risk may be negotiated.
(b) All relevant information available to the transferor ought to be supplied to the recipient of the risk.
(c) The basis of pricing should be on broad categories of risk, rather than related to small details.
(d) Risks which cannot be quantified with sufficient certainty ought not to be priced.

It must be remembered that not all construction project risks can be transferred to third parties. Where a particular project risk cannot be transferred, there are two possible courses of action. The client can either avoid the activity with which the risk is associated or retain the activity but take steps to reduce the risk.

3. *Risk allocation:* This is associated with allocation of risk factors to contracting parties. In the construction industry, construction clients are noted for their unwillingness to take on project risks and, therefore, use their contractual strength to allocate them to contractors/sub-contractors whenever possible (see Figure 3.1). Normally, parties to whom risk has been allocated respond by including an appropriate risk allowance in their tenders. The full extent of these risks allowance are unknown to clients, however, as contractors

normally relate them to the magnitude of risks they are assuming. Usually, the willingness of a contracting party to assume risk is dependent on:

(a) Magnitude of risk and its impact on the construction project.
(b) Ability to bear the consequences of a risk eventuating.
(c) Ability to manage the risk to advantage.
(d) Extent of benefit/return accruing from the risk assumption.
(e) Ability to mitigate the risk factor.

4. *Risk reduction:* Where the client decides to take on the risk factor(s), measures should be adopted to reduce its effects. Contingency plans should also be established to deal with the risks should they occur. This contingency plan may take the form of a percentage allowance added to the project estimate.
5. *Risk removal:* Where the level of risk is deemed uneconomical to accept, the best approach to dealing with it is to remove the activity with which the risk is associated. This may involve reassessment of project strategies, development of alternative design solution or, where necessary, redesign of the project.

3.4 TYPES OF PROJECT SUITABLE FOR APPLICATION

Project risk analysis and management can be used in any type of construction project since all projects contain elements of risk, but it is more beneficial to some projects than to others. For example, on small projects, budget constraints may only justify a low level of application; however, its application is normally beneficial to projects that respond to the following descriptions:

- New technology projects.
- Large capital outlay or investment projects.
- Fast track projects.
- Large refurbishment projects.
- Projects requiring unusual agreements (legal, insurance or contractual).
- Politically or environmentally sensitive projects.
- Projects with tight financial/economic parameters.

3.5 RISK AND CONSTRUCTION PROCESS

As no construction project is risk free, all construction activities expose contracting parties to risk. For this reason, when risks have been identified they should be quantified and formally recognised in the contract documentation (i.e. the project brief, specifications, drawings, tender packages, various contracts and agreements). This measure seeks to formalise equitably any risk in the construction process to enable parties to perform their contractual duties. Moreover, there has to be a clear understanding of what is being bought through the contract and, hence, the risks must be clearly identified and their management, minimisation, transference, sharing and acceptance planned.

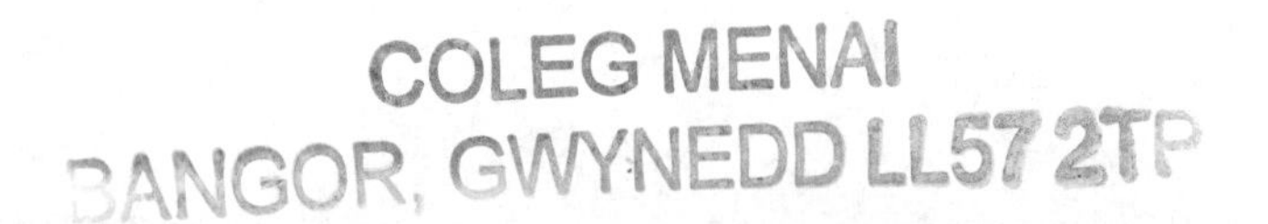

3.6 UNCERTAINTY

Unlike risk, uncertainty is an unknown situation. By its very nature, its possible outcome cannot be analysed or predicted and hence it cannot be transferred to a third party. As previously mentioned, the execution of a construction project can be a complex and difficult process. It is often compounded by the presence of constraints on time, resources and performance and is frequently exacerbated by conflicting objectives of parties involved. This makes the construction project a fertile ground for various forms of uncertainties which include those listed below:

1. *Task uncertainty:* As construction products are unique, the production of almost each one of them calls for architectural/engineering design work and the organisation of the various tasks within the production process. These activities create several problems which delay the production cycle on short-term projects. Typically, however, the unique nature of construction projects and mobile workforces do not make all expertise gained on one project transferable to the next. Each construction project, therefore, has new tasks and associated problems to be solved. The inability to foresee all problems associated with the ever-changing construction tasks and to provide for them beforehand creates a task uncertainty.
2. *Market uncertainty:* The market for construction projects is uncertain, static and the competition is fierce. For most building contractors, all they are certain of is the project in hand. They do not know what their next project will be; they have no prior knowledge of its size, value, location, construction period, the technologies and resources that may be required and so on. As competition heats up, so is the uncertainty of the profit markup that will bring the next project to the contractor.
3. *Weather uncertainty:* Almost all construction activities take place in an unfavourable and uncontrollable external environment. The weather can therefore have a major impact on productivity. Although no one can predict the weather one year hence with precision, the contractor's problem may be exacerbated by an inability to obtain an area's complete geological information needed for weather forecasting. The contractor's vulnerability to the disruptive effects of the weather creates an uncertainty in the management of construction production.
4. *Organisation uncertainty:* The initiation, design and production of construction products is not undertaken by the same group of people (i.e. client professional advisers, contractor's management team and operatives) from project to project. Therefore, there is a need for the establishment of a temporary organisation for every project. In any event, every temporary organisation exchanges influences, overcomes tensions and gains the trust of each other before they can work effectively together as a project team. Such initial tensions and instructions that every project organisation invokes exposes all the members of the project organisation to a state of uncertainty.

5. *Cost-related uncertainty:* In construction projects, the cost of construction is determined before production (i.e. by either negotiations or competitive tendering). Both approaches to price determination apply the principles of estimating for building work in one form or another. But estimating, it must be remembered, is not an exact science and errors in estimating are not uncommon in construction contracts. Therefore, inadequate estimating procedures, errors and contractor's inexperience can make the relationship between the estimated cost and the actual cost become the source of uncertainty in the minds of building contractors.
6. *Resource uncertainty:* Good contract managers or site managers and skilled operatives can, at times (especially when the industry is working under capacity), be difficult to recruit or replace. During such times these personnel can only be recruited at prices higher than those allowed for in the contractor's tender. The same can be said for shortage of materials and/or late deliveries. The effect these shortages can have on the progress and the profitability of a construction project is a cause of uncertainty to building contractors.
7. *Sub-contractor's performance uncertainty:* Construction project contains several specialist trades and activities which are beyond the capabilities of any single individual to master. For this reason, sub-contractors are engaged to provide these specialist services. However, due to a contractor's inability to specify adequately each sub-contractor's working conditions and circumstances, progress on site can be disrupted. The delay arising from disruptions, and the possible overrun on the project programme that this can cause, exposes contractors to a state of uncertainty.
8. *Litigation uncertainty:* The majority of construction contracts suffer litigation stemming from disputes with the client or sub-contractor over several issues, including disruption to the progress of the works. While it is recognised that, in an ideal situation, it is essential to identify each matter of disruption before contract and take the appropriate action to avoid it, in a construction contract this is not feasible. The inability to know, beforehand, the causes of construction disputes and the effect of their resolution is the cause of uncertainty among project participants.
9. *Interest rates/inflation uncertainty:* The lengthy production period for construction projects allow them to become susceptible to upward movement in interest rates/charges and/or inflationary pressures. It is, however, impossible to predict accurately the performance of national and international economies or their influence on interest rates and inflation. The inability to forecast the interest rates and inflationary pressures and allow adequately for them in tenders is a source of uncertainty among building contractors.
10. *Contracting uncertainty:* The production process in construction projects is complex with many hidden variables, and neither the building contractor nor the client's professional advisers are able to judge accurately in advance exactly what to expect. Consequently, there are a host of unknowns which come to

light when site production starts; for example, the ground condition may be different from that envisaged and may cause problems during excavation and/ or substructure works. This state of affair is the cause of uncertainty among project participants.

11. *Letting or sale uncertainty:* Where the construction project is intended for letting or sale, the lengthy construction period creates uncertainty as to whether there will be a market for the product when completed. By the time the project is completed several changes in the marketplace, such as a fall in demand and/or price, rent controls, higher tax liabilities and so on, may have taken place making the project unprofitable.

The above list, although not exhaustive, shows that there are many uncertainties associated with the construction process and, moreover, reinforces the fact that construction is a complex and risky business. The immensity of these uncertainties requires skill and experience to identify them all. However, in spite of this, it is the characteristic of construction projects to often proceed despite high degrees of uncertainty. The participants often display remarkable commercial willingness to go forward with uncertainty as to design solutions, uncertainty as to the eventual scope of the works and uncertainty as to the time periods that will realistically be required in order to complete the works. Traditionally, the need to address some of these uncertainties is postponed, leaving enormous discretion to be exercised in the future by those administering the contract of the post-contract phase. But by realistic appraisal of levels of uncertainty at the outset, it is often possible to remove the seeds of what would otherwise be an eventual problem. For this reason, it is advisable to seek the advice of experts who are able to analyse uncertain events (using, for instance, sensitivity analysis, cost–benefit analysis and probability theories), and are skilful in removing events from areas of uncertainty to areas of risk and/or certainty. It is only through the guidance of these experts that an adequate provision can be made for the eventuality of an uncertain event.

3.7 INFORMATION COMMUNICATION MANAGEMENT

As mentioned above, construction projects are complex and risky and therefore require active participation of all parties who must be kept informed of the project's requirements at all times. Hence, information is a prime source of activity in the construction process. It rarely has a simple, clearly defined starting point because, strictly speaking, the producer of information usually builds on and adds to data received from others. By and large, steady progress on a construction project depends on the right party obtaining the right information at the right time. Accordingly, construction project participants cannot perform effectively without an adequate accurate, and timely flow of information. For this reason, each participant in the construction process has responsibility for transmitting information and communication as:

- Non-receipt or late receipt of construction information is the cause of contractual claims and disputes.
- Late information means that remedies for construction problems become more expensive.

Also, the nature, volume, direction and timing of the flow of construction information all vary considerably and, hence, this demands its effective coordination, control and dissemination to ensure their proper utilisation. The ultimate aim of coordination is the integration of ideas and activities, the preservation of unity of objective and the fostering of cooperation between project participants. Nevertheless, the foregoing aims may be difficult to achieve without an effective communication process and, for that reason, it can be said that coordination problems are essentially those of communication. Consequently, it is essential that formal communication links are established between the design team members soon after they are appointed (Figure 3.2).

It is also essential that a logically organised information system capable of providing each participant with a flow of understandable construction information is established before commencement of design. Moreover, it is vital to coordinate the variety of project information as and when produced in order that the following distinct but interrelated activities can be undertaken effectively.

- Establishment of size, specification and cost parameters of project.
- Means of achieving adequacy of design interface between project participants and enhancement of efficiency of design.
- Selection of an appropriate contract and procurement method in line with project needs.

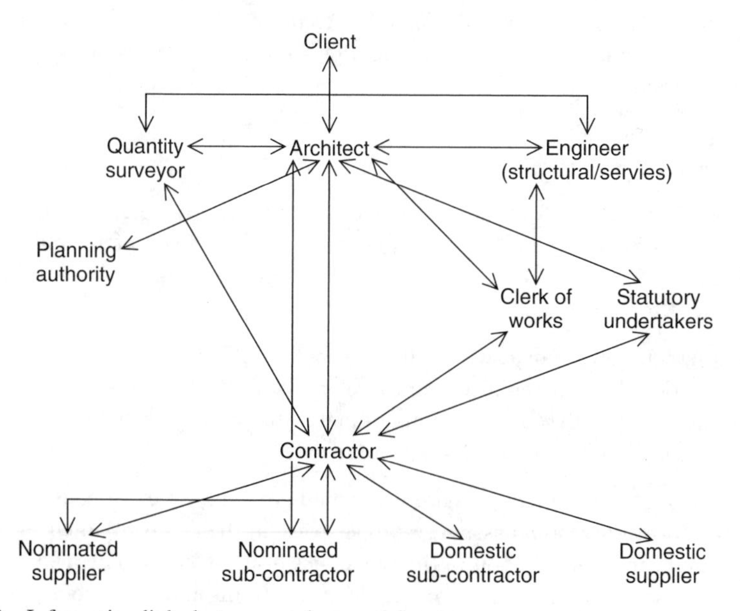

Fig. 3.2 Information links between project participants

- Programming work to delineate start and duration constraints.
- Production of adequate and clear contract documentation which avoids conflicts and disputes.
- Commissioning building and services to meet client's exacting expectation, user and performance requirement.
- Enhancement of general creative decision - making in all stages of building production and management.

3.7.1 Production and use of project information

All construction information produced, utilised and shared among project participants is intended to contribute to the success of a particular project. Under the traditional procurement method (where design and construction functions are separated), the main producers of construction information and stages of production may be summarised as in Table 3.1. Furthermore, this table demonstrates that construction information is produced and utilised in all the design and production phases by the project participants. For example, the quantity surveyor produces cost estimates and bills of quantities during the design phase for the contractor, architect and client's use; and produces cost report and final account also for their use at the production phase.

3.7.2 Intra- and inter-organisation coordination

Construction activity, unless designed and constructed by the same company, is undertaken by specialist participants from several establishments organised into a temporary group over an agreed timescale. Accordingly, there is a need for coordination of information at intra- and inter-organisation levels to ensure the success of the project.

Intra-organisation coordination

To function effectively and remain in business, managers of companies must coordinate the activities of their personnel for the achievement of a common specific goal. Above all, there is also a need to exercise coordination of construction information to ensure that:

- Information flow is adequate, continuous and uninterrupted.
- Information flow is simplified, improved and controlled.
- Information is analysed and communicated to the right personnel.
- Information is released at the appropriate time.

Furthermore, the production, coordination and communication of construction project information should assume a proper place in the company to enable it to make an effective contribution to the economies of production. The neglect of intra-organisation information coordination could result in one or more of the following:

Table 3.1 Production and use of project information

Information producer	Project phase	Form of information	Information user	Remarks
Client	Conception	Brief	Architect	The ultimate users of the finished product may be interested in the information
Architect	Feasibility	Report	Client	Drawings may be for various design solutions
	Design	Drawings	Client, QS, engineer and contractor	
	Construction	Revised drawings		
	Commissioning	As-built drawings	Client	
Engineer (Structural/ Services)	Design	Drawings	Architect, QS, contractors	
	Construction	Revised drawings		
	Commissioning	Operating/ Maintenance Manual	Client/ ultimate users	The commissioning input is normally from the services engineer
Quantity Surveyor	Design	Cost estimates and Bills of quantities	Architect, client	
	Construction	Cost report Final Account	Architect, client, contractor	Provides cost advice (comparable estimate) prior to receipt of tenders
Contractor	Construction	Progress report Contractual claim	Architect, QS, client	Ultimate user of the finished product may be interested in the information
	Commissioning	Maintenance manual	Architect, client	
Material suppliers	Construction	Material availability and supply report	Contractor	For nominated supplier, the architect may be interested in the information

- Lack of cross-fertilisation of ideas between construction projects and clients.
- Absence of feedback on completed construction projects and perpetuation of inherent defects and expensive mistakes from one project to another.
- Deficiency of knowledge on company policies and practices.
- Abortive work and/or duplication of efforts.
- Failure to communicate essential construction information to the right person at the right time.

- Conflict of construction information, such as that shown on the architect's and engineer's drawings.

Inter-organisation coordination

The temporary organisation structure formed for the execution of a particular construction project consists of specialist participants from different establishments and, therefore, demands the effective coordination of information to ensure that the efforts of all participants are directed to the needs of the project. A lack of coordination in design, for example, results in one or more of the following:

- Discrepancies between architect's and engineer's drawings (e.g. dimensional, shape, details variance).
- Conflict of architect/engineer's drawings with those of services engineer (e.g. position and size of holes, ducts).
- Conflict of services engineer's drawings with those of another services engineer (e.g. position of holes, ducts, equipment).
- Architect/engineer unaware of the level of details or construction information required by the site production staff.
- Improper channels for distribution of construction information.

In addition, the following are the consequences of uncoordinated construction project information:

- Discrepancies between item descriptions in bills of quantities and details shown on architect/engineer's drawings.
- Inclusion of inappropriate and/or out-of-date information in contract documentation.
- Contract documents which disagree with each other by reason of inconsistencies, ambiguities and/or omissions.
- Late issue of information (e.g. late arrival of architect/engineer's drawings on-site).
- Hastily produced construction information (e.g. drawings hastily produced and, hence, devoid of the required information).
- Site management (contract managers, site agents, foremen and clerk of works) spending much of their time sorting out discrepancies and/or ambiguities in contract documentation. This robs them of valuable time that could otherwise have been used more beneficially in supervising/monitoring the quality aspects of the works.
- Uncontrollable variations, delays, disputes, increased cost, claims and consequential uncertainty as to the final cost of the construction project.
- Completed product of poor quality with technical defects and the resultant increased occupancy cost.

SUMMARY

Construction projects are complex and are bounded by risks and uncertainties. There are, therefore, various elements of it that may be subject to risk or uncertainty. For example, there could be delays in production, rises in production costs, changes in the cost of borrowing, problems of letting the completed facility, shifts in market conditions or a whole host of other changes which may later affect the project. For this reason, prior to the decision to build, there is a need to analyse the effects of the project risks and uncertainties and, where necessary, provide adequately for them. But while project risks can be analysed and quantified and also possibly managed, project uncertainties do not fall easily into this equation and, therefore, require more attention. Moreover, the success of construction project risk management requires the input of all the project participants who must be in constant touch with each other and spread project information through an effective communication medium which should be well established and managed.

FURTHER READING

The Association of Project Managers 1992 *Project risk analysis and management*. APM.

Crawshaw, D.T. 1979 Project information at the pre-construction stage. *BRE Paper IP 27/79.*

Fellow, R.F. 1989 The management of risk. *CIOB Technical Information Service, No 111.*

Groak, S.; Householder, J. 1992 Contractor's uncertainties and client's intervention. *Habitat International*, **4** (2/3), pp. 119–25.

Norris, C.; Perry, J.; Simon, P. 1992 *Project risk analysis and management*. The Association of Project Managers.

Patterson, J. 1977 *Information methods for design and construction*. John Wiley & Sons.

Toakley, A.R. 1990/1991 The nature of risk and uncertainty in the building procurement process. *Australian Institute of Building Papers, No 4.*

Uler, T. 1992 Risk management in construction. *The Chartered Builder (Australia)*, February, pp. 21, 23–4.

Ward, S.C.; Chapman, CB.; Curtis, B. 1991 On the allocation of risk in construction projects *International Journal of Project Management*, **9** (3), August.

Wearne, S.H. 1992 Contract administration and project risks. *International Journal of Project Management*, **10** (1), February.

Yates, A. 1986 Assessing uncertainty. *Chartered Quantity Surveyor*, November, p. 27.

CHAPTER 4

INITIATION OF CONSTRUCTION PROCESS

4.1 INTRODUCTION

Having decided to undertake a construction project, the client's needs must be met to his or her satisfaction. However, every construction process is a complex and high-risk venture and, moreover, a construction project is a unique undertaking subject to pressures of fluctuating resources, procurement times, supply problems, weather conditions and so on. Accordingly, a number of construction professionals offer varied services to assist in the realisation of clients' development objectives. These professionals attend to matters which include the determination of exactly what is required by the client. Usually, the client's requirement includes the design of the product, its production within quality, cost and time parameters and in conformity to health and safety and other statutory regulations. The extent and timing of the procurement of professional services vary from client to client and also from project to project; however, in most cases, the services of construction professionals become necessary from inception of the project. On appointment, the construction professionals are briefed by the client and they in turn follow this up with planning, designing and site production management.

4.2 PROJECT BRIEF

Briefing is a process where the client defines his or her construction project requirements and communicates them to an architect of his or her choice. The client's initial brief is a statement of intent which will normally include value for money; a construction product that is pleasing to look at, is fit for its purpose and is delivered on time, is free from defects on completion, has reasonable running costs and satisfactory durability.

As the above may be inadequate for proper interpretation, following the initial brief, the architect arranges a series of exploratory meetings for the identification and clarification of the following important points:

- The type of construction project being proposed.
- Aims, resources and context.
- Design requirements.

These exploratory meetings assist the architect in the identification of services required, outline problems and their likely solutions. It also averts a situation whereby the client subsequently requires detailed changes in the project, with serious implications on programme and cost. These meetings also outline the priorities and set out the *critical* aspects of the project (i.e. aspects which will cause the project to be aborted if not fulfilled). Therefore, normally before the development of the project brief, the architect seeks further information on a number of points in connection with the client's development proposal. These include the following:

- The status of the client.
- General nature of the client's requirement.
- An indication of project's timescale.
- Details of land and legal constraints (such as covenants and easements), and whether owned.
- An indication of cost limit.
- Client's financial resources to complete the proposed construction project.
- Status of outline planning consent.
- Client's technical contribution (if any) to the proposed construction project.
- Other consultants (if any) who have been involved in the proposed construction project.
- Details of architect's appointment and whether in competition with other firms or designers.

The client's responses to and/or comments on the above points, and many more questions and answers thereafter, would enable the architect to:

- Understand the client's expectation and, hence, provide technical interpretation of the brief and commence its development if necessary.
- Specify the quality required in terms of effects of the client and the public's use.
- Resolve any conflicts of interest and set out all the possibilities.
- Set out methods and timing of any sources of finance, tax allowances, grants which have a direct influence on the commencement of design and construction of the project.
- Identify all or most of the external factors that may influence the time, cost and quality of the proposed construction project (e.g. planning permission and local regulations).
- Balance objectives, resources and context which are clear enough and permit design so that consultants can use their skills and knowledge in these fields.

During the briefing stage, a client may find it helpful to involve prospective or ultimate occupants or users of the completed facility in the development of the brief so that the objective of the project may be achieved. In addition, where different parts of a client's establishment have conflicting priorities on cost, time and design, the initial project brief may highlight these disagreements. The architect will include the need for solution of this incompatible requirement in the development of the brief. Furthermore, during the exploratory meetings with the client, the architect will normally advise on the appointment of other consultants (the quantity surveyor, the structural and services engineers) and clarify services to be provided, conditions and basis for fees of all consultants with the client. When the client agrees to employ the services of the consultants, the architect advises the client to confirm their appointment in writing and, on confirmation, the consultants at this phase of the confirmation process become known as the *design team* and the architect acts as a design team leader. This set of relationship serves to indicate the need for a team effort when formulating the development proposal. The team should include a valuer who has the expertise to advise on the best type of development and its ultimate value. He or she can also give useful information on the most suitable type of construction and quality of finish.

4.2.1 Attributes of a good brief

The success of all construction projects depends primarily on a good brief, and a good brief is the one which is adequate in terms of:

1. *Clarity:* The purposes of the brief should be clear, accurate and capable of detailed development so that due consideration can be given to each aspect of the construction project.
2. *Priorities:* The brief should establish the project's priorities and define those particular items/requirements that are necessities and those that are optional wishes.
3. *Consistency:* The brief should be consistent throughout, and devoid of conflicting statements which frustrate the determination of construction project design and cost.
4. *Completeness:* At any stage in the development of the brief, it should be complete enabling production of a building that satisfies the client's needs.
5. *Realism:* The brief should be realistic and, hence, shaped to a greater or lesser extent by resources available and the quality to be achieved.
6. *Relevance:* The brief should contain the relevant information and decisions essential to the successful realisation of the client's development objective.
7. *Benefits of development:* The brief should establish the benefits the client hopes to obtain from the proposed construction project and the means of achieving them.
8. *Flexibility:* The brief should be broad, flexible and should also lend itself to exploration of design problems, options and uncertainties.

Functions of the brief

Once the client's brief has been developed, agreed and formalised as a document, it can serve various functions, including:

- A key working document setting out the client's requirements in terms of net-use floor areas; specific details of the various space functions, their relationships, particular matters to do with constraints; and details of the involvement of other parties.
- A channel of communication conveying instructions, decisions and information between a client and his or her project team.
- A means of stimulating communication and discussion among project participants by facilitating the setting of priorities, analysis, problem identification and information flow. It also provides a collective 'thinking through' of the project.
- A record of decisions, instructions, information, agreement, amendments, conflicts and uncertainties which can be used to ensure continuity and consistency during the construction process, or as a reference document when the completed facility is in use.
- A tool for evaluation of proposed construction projects aims and resources against which design contents and framework can be appraised.
- A basis for estimating resources required, giving information of general comfort standards, cost, timescale for design/production of construction product and estimated useful life.
- A contractual document accompanying a legal agreement between the client and the design team.

For many construction projects, briefs are most useful initially for promoting discussion and securing the involvement of all project participants. As the brief and design develop, the brief becomes most useful as a documentary record since, by the time the brief is fully developed, the design will also be largely formed.

4.3 PLANNING GENERALLY

Construction projects, like all projects, usually involve combining the efforts of a number of people to achieve some set objective against a fixed timescale. To achieve the set objective defined by the client (i.e. delivery of a product of desired quality at an acceptable price within required time and safety), the project participants undertake planning. Generally, planning is about thinking ahead and depends upon the existence of alternatives and then decisions made regarding what to do, how to do it and by whom it must be done. Usually, the more people involved in the planning process the better is the outcome. But, the planners should realise that an important starting point for planning must be to do everything to reduce uncertainty. In addition, a plan should form the basis for the direction and control of a construction project. Moreover, it provides a discipline, forcing project participants to take regular, careful forward predictions and hence requires rigorous communication about goals, constraints and strategic issues.

4.3.1 Construction project planning

Once a client has appointed the construction professionals, their immediate action is to get together to formulate the project plan. In the process, meetings are organised in which questions about the project *how*, *when*, *where*, *who and with what* are raised and discussed. The planning must be based on a clear division of the construction project into stages to facilitate manageability and control; and the degree of detail put into the planning depends on various factors, among them:

- Size of the construction project.
- Degree of difficulty.
- Degree and extent of indeterminate operations.
- Knowledge and experience of operations.
- Knowledge and disposition of project participants.
- Extent of internal and external constraints.

A graphical schedule known as a *programme* forms the basis for effective planning. All good programmes show sufficient detail which enables proper consideration to be given to timing and duration of various stages of the project. Also, it provides the project participants with a yardstick against which to measure progress and a basis for regular review. At the early stages of a construction project, a simple bar chart (see Figure 4.1), setting a timescale for the various key stages of the project, serves as an important common reference tool for strategic decision making for all project participants. The programme would be prepared on the assumption that all design information would be available when required. But, as the construction process does not normally take place in an ideal situation, this is rarely the case. Moreover, the success or failure of all projects depends on work done at this stage. Therefore, more attention and time should be given to the construction project at this stage to facilitate quick production of construction information as and when required at the production phase. There will also be a need to produce and issue reports on all decisions taken at these initial project meetings so that the proposed timing, methods and strategy available are understood by all participants.

4.4 FEASIBILITY OF PROJECT

Once a project brief has been formulated, the design team can proceed with the assessment of its feasibility. This process, known as a *feasibility study*, is undertaken to examine the client's construction project proposals in order to establish the degree to which it is practicable. A practicable project is one which is manageable and achievable, within a client's budget, on time, economically and safely. During the feasibility stage, the client should be in a position to provide the design team with all the required relevant project information and any necessary assistance. The study will determine the feasibility of the client's requirements and may involve a review with the client's alternative design, production approaches and cost implications. Other matters of importance that may be considered at this stage are

Fig. 4.1 Pre-contract programme

Item	June	July	Aug.	Sept.	Oct.	Nov.	Dec.	Jan.
Stage								
Feasibility study/report								
Outline proposals								
Planning applications								
Scheme design								
Detail design								
Consultants drawings								
Production information								
Statutory approvals								
Bills of quantities								
Printing of bills & tender documentation								
Tender action								
Analyse/report on tenders								
Pre-contract planning								

programme, advice on planning permission, approvals under building regulations and other similar statutory requirements.

Feasibility studies thus seek to establish whether or not a given proposal is worth while, taking into account all constraints. For example, before decisions can be taken by management concerning the siting, design and construction or adaptation of buildings, they would need to investigate the choices and options available together with the likely consequences of actions selected. Also, a framework will be needed for the project's participants to enable them to find some common collective understanding of the reasoning behind various decisions taken. In this way, all those involved or affected by a client's development proposal may contribute towards influencing the final outcome.

Functions of feasibility studies

Once a feasibility study has been completed it can be put to many uses, including:

- Facilitating detailed examination of a client's brief and, hence, prompting discussions of objectives, costs and implications of the development proposal. Such a dialogue among construction project participants highlights any conflicting objectives stated in the original brief.
- Easy systematic examination of a short-listed set of options and, hence, minimising the risk of an omission of an important factor from the assessment and comparison of various alternatives.
- Pointing out which factors of the client's construction project are critical to its success (e.g. the timing of the project, interest rates, the rate of inflation, other developments taking place and so on).
- Establishment, in advance of any major expense and commitment, the viability and profitability of the client's construction project taking into account the capital and running costs, timing and rate of return.

4.4.1 Feasibility report

On completion of the feasibility studies, the design team prepares a report designed to highlight, evaluate and structure the advantages and disadvantages over time of the alternative solutions to the client's development proposal. Generally, the contents of feasibility reports vary in accordance with the type of construction project, type of client and his or her requirements. However, a typical report should contain some of the following issues:

- Brief – some clients find it difficult to formulate their brief without professional assistance. Therefore, the final agreed brief as understood by the design team must be spelled out to avoid any future misunderstanding.
- Suitability of site for the construction project proposed – its location, access limitations, topography, nature of ground and its maximum safe-bearing capacity.
- Existing infrastructure such as access roads, public sewer, electricity, gas, water and so on in the vicinity of the site.

- Costs – the target cost of the proposed construction project.
- Accommodation – the type of accommodation and a list of total floor areas that can be provided within the target cost.
- Statutory regulations and requirements (e.g. planning, legislation and building regulation controls).
- Environmental considerations and the measures to be adopted to reduce its impact on the construction project, and vice versa.
- Sketch of scheme – architect's sketches of preliminary elevations or sketch perspective of the completed design, the presentation of which should be readily understandable by a lay client examining the scheme.

As feasibility studies provide an objective approach to evaluating competing proposals, the feasibility report will also recommend a course of action which may induce the client to abandon, modify or continue the project. At this point the client should be able to decide whether or not the project should go ahead and which sketch options appear to meet his or her other desires.

4.5 PROJECT APPRAISAL

Besides the desire to know whether planning permission will be granted for the development on the selected plot of land, a client would also like the financial viability of his or her development project to be appraised. The appraisal normally involves the collection and study of data such as the probable value, revenue and cost of the proposed construction project (see Figure 4.2). Such a study enables the client's professional advisers to produce a developer's budget, which highlights the total cost and the financial return to be expected from the property investment. Further, when a client is developing for sale, letting or leasing, besides information on production cost that he or she may require from the quantity surveyor, he or she may also require information on land, probable value and revenue on completion. This information is in the domain of the valuer whose services are procured at this stage of the construction development.

4.6 DEVELOPER'S BUDGET

As part of a development of the feasibility studies and project appraisal, the client will need to know, at the initial stages of his or her development aims, the total cost of the project. Accordingly, the client's professional advisers normally undertake a cost exercise known as *developer's budget* which provides the client with information on the following:

1. *Gross development value:* The gross development value (GDV) is the estimated total rent per annum accruing from the completed development less reasonable allowance for outgoings (e.g. maintenance, repairs, management and so on).

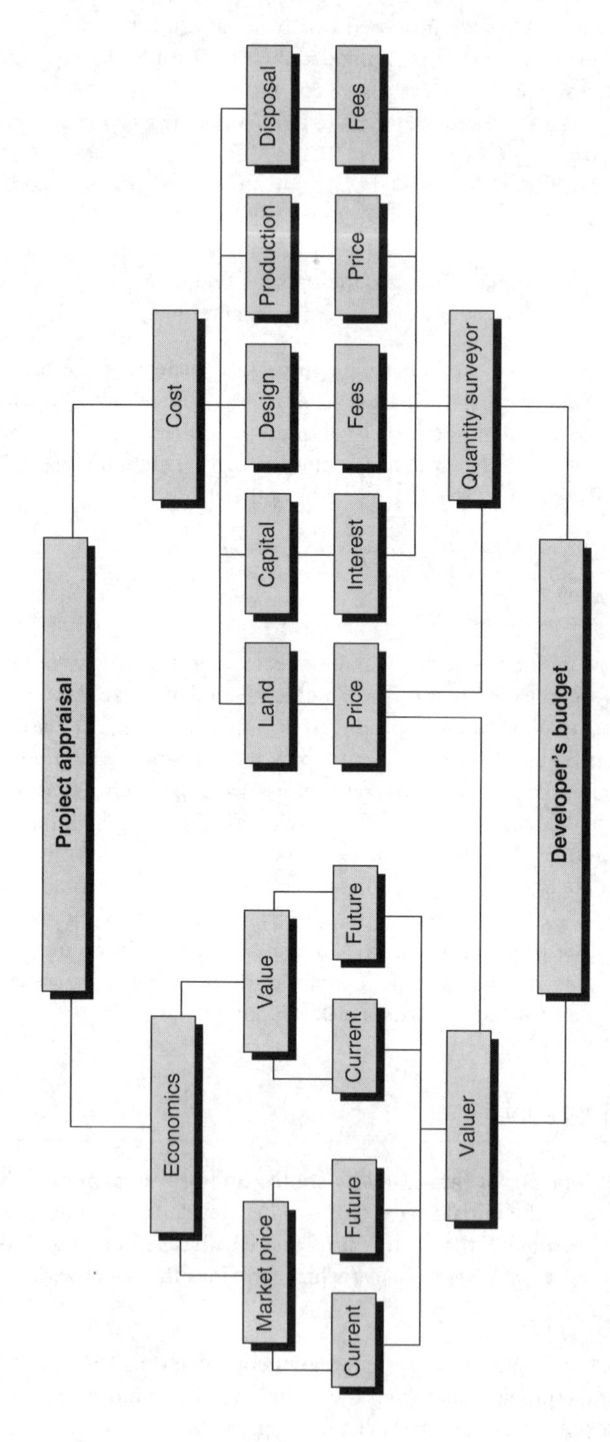

Fig. 4.2 Components of the client's project appraisal

The balance then provides the net income which is capitalised by the use of a multiplier commonly known as *year's purchase* (usually based on recent sales in the area) and the total obtained is termed the *gross development value* (GDV).

2. *Total cost of development:* This comprises the sum of land cost, total construction cost, finance charges and fees paid by the client in return for the completed development.
3. *Total construction cost:* The total construction cost comprises cost of the new building work, services and the cost of site works such as landscaping, roads, drainage and other ancillary site services, which are priced at current unit *all in* rates. The resultant figure represents the net cost of the fabric of the building, to which needs to be added drainage and external works and cost of site preparation. The estimated cost may be prepared on an elemental basis, using historical cost data, possibly form a number of buildings, adjusted to the requirement of the new building (see Example 4.1). Alternatively, where adequate project information is available, approximate quantities may be taken and used to arrive at the estimated total cost of the project.
4. *Maximum permissible expenditure on land:* The client may decide to purchase the land in advance of the financial feasibility exercise, in which case the cost of land must include the cost of physical preparation of the site such as clearing existing structures, obstructions and trees before the commencement of new construction works. On the other hand, the client may delay the purchase of land until the financial feasibility exercise has established the maximum amount he or she can spend on the land purchase transaction (see Example 4.1).
5. *Total fees:* This is the amount of fees the client pays to his or her construction professional advisers, legal advisers, estate agents, and so on. The extent of fees generally depends on the level of service required.
6. *Total cost of finance:* This is the total interest on capital borrowed and used in land purchase and carrying out the construction work. The calculation of interest on money borrowed for site purchase is usually based on the period of actual purchase until the development is sold or let and the debt paid off. However, the construction cost is met monthly through interim valuations and hence the full cost of financing will not be met at the beginning of the period of construction. For this reason, the cost of finance for construction works is calculated at an agreed rate of interest on half the building cost for the entire duration of the construction contract (i.e. the average cost over the construction period).
7. *Developers profit:* This is the financial return the client expects from the development with an allowance for the risks involved in undertaking the development. The rate varies according to the type of development and the likely degree of risks, such as the inability to lease, falling rent, high interest rates, inflation and generally the volatile property market to be expected. An allowance of 10–20 per cent of gross development value is normally considered to be a reasonable margin of profit.

Example 4.1 Preliminary estimate of cost

Project: UPPER DAGNALL STREET DEVELOPMENT
Date of estimate: Date of tender:

	£	£
Gross floor area: 2500 m^2		
Basis of estimate:	£ per m^2	
Office Block, London (Date) 1750 m^2	850.00	
Adjustments		
1. Market conditions		
RICS cost information index: *Add* 5%	42.50	
	892.50	
2. Total height of building		
6 storey and basement in lieu of 6 without basement		
Add 3%	26.78	
3. Site conditions and influence on foundations	919.28	
Increased number diameter and depth of piling: *Add*	17.52	
	936.80	
4. Specification level		
High quality curtain walling (see notes): *Add*	10.25	
	947.05	
5. Inclusions and exclusions		
(a) Shell and core approach; (b) Plant room in basement in lieu of roof } *Deduct*	14.01	
	933.04	
6. Services		
See services engineer's information: *Deduct*	5.50	
	927.54	
7. Other factors		
Tendering; Design risk } *Add* $3\frac{1}{2}$%	32.46	
	£960.00	
Total estimated building cost: 2500 m^2 @ £960/m^2		£2,400,000.00
Add		
(a) Drainage and external works		250,000.00
(b) Cost of site preparation		50,000.00
Total estimated cost of the development		**£2,700,000.00**

Note: Professional fees, cost of financing and VAT excluded

A developer's budget in profile

Let us assume that a hypothetical client has found a freehold prime site which has a frontage of 50 metres, a depth of 45 metres and is suitable for offices and shop development. The client's professional advisers' initial action is to find out from the area planning authority if planning permission will be granted for the proposed development project. If planning permission can be obtained, the client will then seek professional advice on the following points:

- Permitted size of development for the site and design proposals.
- An indication of probable development cost.
- Likely rental income or selling price to be expected for the property in that location.
- Feasibility of the development in technical terms.
- Maximum price that he or she should pay for the plot of land.
- Financial viability of the proposed development project.

The client's professional advisers may be able to respond to the client's initial development enquiry after the preparation of the following development budgets, which will determine:

- Maximum amount that can be spent on the purchase of the land (i.e. site value).
- Level of profit the client should expect from the development.

Worked examples 4.2 and 4.3 are based on the following data provided by the client and the client's professional advisers.

Data

- Area of site: 50 x 45 metres.
- Total floor area of development: 2,500 m^2.
- Cost of site preparation: £50,000.
- Construction costs: £960/m^2.
- Site works: £250,000.
- Professional fees: 15 per cent of total development cost.
- Cost of finance: $7\frac{1}{2}$ per cent.
- Legal, agents' fees: $2\frac{1}{2}$ per cent of GDV.
- Estimated disposal income: £380/m^2.
- Year's purchase (YP) in perpetuity @ 8 per cent.
- Gross developer's profit: 15 per cent of GDV.
- Development completed and let or sold: 24 months after site purchase.
- Construction period: 18 months.

Example 4.2 Developer's budget

Development budget to determine how much the client can offer for a site.

	£	£
A. Gross Development Value (GDV)		
- Net lettable space = 2,050 m^2 @ £380/m^2	779,000.00	
- Management @ $2\frac{1}{2}$%	19,475.00	
Net income	759,525.00	
- YP perpetuity 8%	× 12.50	**9,494,062.50**
B. Cost of development		
(1) Site preparation	50,000.00	
(2) Construction costs		
gross floor area – 2,500 m^2 @ £960/m^2	2,400,000.00	
(3) Drainage and external works	250,000.00	
Plus Professional fees	2,700,000.00	
(Architect, quantity surveyor, structural engineer, heating & ventilating engineer) @ 15%	405,000.00	
	3,105,000.00	
(4) Finance costs		
(i) Construction costs @ 8% for half construction period		
$\frac{£3{,}055{,}000}{2} \times 1.5 \times 8\%$	183,300.00	
(ii) Interest on site preparation costs = £50,000 × 2 × 8%	8,000.00	
(5) Disposal costs		
$2\frac{1}{2}$% of GDV = £94,940,625.50 × $2\frac{1}{2}$%	237,351.56	
(6) Contingencies: 5% of construction costs = £3,105,000 × 5%	155,250.00	
Total costs	**3,688,901.56**	
C. Developer's profit: 15% of GDV £9,494,062.50	1,424,109.30	
Total costs	**5,113,010.86**	**5,113,010.86**
Residual amount for site		**4,381,051.64**
Less		
(i) Cost of site finance 8% on, say £3,000,000 for 2 years = £3,000,000 × 2 × 8%	480,000.00	
(ii) 4% for incidental acquisition costs = £3,000,000 × 4%	120,000.00	600,000.00
Site value to client		**3,781,051.64**

Example 4.3 Developer's budget

Development budget to determine client's development profit

	£	£
A. Total cost of land		
(1) Cost of site	200,000.00	
(2) Site Finance for 2 years @ 8% = £200,000 × 2 × 8%	32,000.00	
(3) Incidental costs of purchase (say, 4% = £200,000.00 x 4%)	8,000.00	**240,000.00**
B. Total construction costs		
(1) Cost of site preparation	50,000.00	
(2) Gross floor area: 2,500 m^2 @ £960/m^2	240,000.00	
(3) Drainage and external works	250,000.00	
Plus: Professional fees	2,700,000.00	
(Architect, quantity surveyor, structural engineer, heating/ventilating engineer) @ 15%	405,000.00	
	3,105,000.00	
(4) Finance costs		
(i) Construction costs @ 8% for half construction period $\frac{£3,055,000 \times 1.5 \times 8\%}{2}$	183,000.00	
(ii) Interest on site preparation cost = £50,000 × 2 × 8%	8,000.00	
(5) Contingencies: 5% of construction costs = £3,105,000 × 5%	155,250.00	**3,451,550.00**
C. Disposal costs		
Legal, agency and advertising costs $2\frac{1}{2}$% of construction costs £3,451,550.00		86,238.75
Total development cost		**£3,777,838.75**
D. Estimated disposal income		4,000,788.75
Less: Development costs		3,777,835.75
Development profit (received after twenty-four months)		**£ 224,000.00**

4.7 THE DESIGN

4.7.1 Preliminary estimate

The primary purpose of a preliminary estimate is to provide early indications of the probable cost of a construction project. In other words, it is an attempt to forecast the cost of a project. Although the preliminary estimate is always prepared on scanty construction data, it is one of the important pieces of information that influence the client's decision to engage in a construction project. By this process, the client is made aware of the likely financial commitment prior to extensive design and contract documentation. The preliminary estimate fulfils a secondary function in providing design team members with early cost information which influences design solutions such as forms of construction, level of specification, finishes and so forth. The preliminary estimate is generally prepared on limited information and its accuracy, to a greater extent, depends on the availability of reliable data on which to base calculations and also the skill/experience of the quantity surveyor in charge of its preparation. However, it is important that these estimates are adequately prepared as incorrect preliminary estimates give a project a bad start. Therefore, although, at this point, the project requirements are invariably at their formulation stage, the quantity surveyor must have reasonably precise information on what is to be included in the estimate. In addition, construction items such as floor and wall finishing, partitions, joinery fittings, special plant, external works and the like must be treated fully in the estimate to ensure that they are allocated an adequate financial allowance. Furthermore, it is always difficult to find the right rate for use in the pricing and, hence, a similar job executed in the quantity surveyor's office is the most reliable starting point. However, as this pricing source is historical, its cost data need adjustment to reflect:

- Current market conditions compared to the rates of the completed project being considered.
- Size, plan shape, storey height and similar factors.
- Specification levels.
- Inclusions and exclusions.
- Extent of services.
- Ground conditions and design solution adopted.

An indication of how these various adjustments are made was given in Example 4.1.

4.7.2 Types of preliminary estimates

The preparation of preliminary estimates consists of processes of measurement and pricing but the techniques used in practice vary and may be classified as either single or multi-rate estimates or methods.

Single rate methods

Preliminary estimates which adopt the single rate method in their preparation are:

1. *Unit method:* This is the simplest and quickest estimating method. It consists of selecting a standard unit of accommodation (e.g. standard unit for school per place; hospitals per bed; theatres per seat and so on) and multiplying the number of required units by the estimated costs per unit.

   Certain types of development lend themselves to estimates based on a price per place, seat, bed and the like and the method gives accurate results when used on projects such as schools, car parks, hospitals and theatres.

   The advantages and disadvantages of the use of the unit method may be summarised as follows:

   Advantages
   - The method provides a quick cost comparison between similar schemes of different sizes and in different locations.
   - It provides a convenient form of establishing a cost limit.

   Disadvantages
   - Use is limited to specific buildings where the building's function can be expressed in unit terms.
   - It is impossible to make allowance for variations of size and plan shape, storey height, form of construction, finishings and fittings.

   Applications: The unit method of preliminary estimating is suitable for use at the inception of a construction process when the client may only be able to state his or her requirements in simple terms of, for example, numbers of people or equipment and so on expected to use the completed development.
2. *Superficial or floor area method:* This method involves measurement of the total internal (i.e. within external walls) superficial floor area of buildings and the total area multiplied by a calculated unit rate per square metre. This method is widely used due to the fact that areas of a building and cost expressed in this way are easily understood by most clients.

   The advantages and disadvantages of the use of the superficial or floor area method may be summarised as follows:

   Advantages
   - The method provides a quick method of preparing preliminary estimates.
   - It is fairly readily understood by clients as most express their requirements in floor areas.
   - There is a wealth of published information available on costs of various buildings for this method of preliminary estimating.
   - The designer can use the method as a guide to cost when preparing sketch designs.
   - It allows cost comparisons to be made between buildings of different sizes.

Disadvantages

- It is rather difficult to make adjustments to a cost per square metre to allow for differences in plan shape or storey heights.

Application: The superficial or floor area method of preliminary estimating is suitable to use during the consultation and feasibility stages of a construction project.

3. *Cubic method:* This is a preliminary estimating method which is now rarely used in practice. In its simple terms, the calculated cost per cubic metre of completed projects are used to calculate the preliminary estimate of proposed development projects. The method has fallen into disuse and its main disadvantage is that the cubic metre rate cannot be adjusted for changes in size, plan shape and storey height of buildings.
4. *Storey enclosure method:* This method of preliminary estimating is claimed to be more accurate than any of the single rate methods of preliminary estimating considered above. The method consists of measuring areas of the vertical and horizontal planes of the building, weighting the planes according to their position in the building and multiplying the total so obtained by a calculated unit rate.

Rules of measurement for storey enclosure method

- Measure internal area of lowest floor and multiply by 2 that part which is above ground level and multiply by 3 that part (if any) which is below ground level.
- Measure the area of roof (measured flat in all cases including eaves).
- Measure the area of each upper floor and multiply by 2; further adjust by multiplying by a progressive factor of 0.15 (e.g. first floor multiplied by 2.15, second floor multiplied by 2.30, third floor multiplied by 2.45 and so on).
- Measure the area of the external enclosing walls. Multiply by 2 the area of only enclosing walls below ground level.

The advantages and disadvantages of the use of the storey enclosure method may be summarised as follows:

Advantages:

- It is a quick way of preparing a reliable preliminary estimate.
- The method takes into account many of the cost factors which other single rate methods of estimating ignore.
- It takes into account the varying cost of elements located in different areas of a building.

Disadvantages

- The subdivision of the estimate into elements or individual items is not practicable.
- The method excludes the value of items such as mechanical installation which in these days accounts for a substantial proportion of a total cost of construction projects.
- The unit rates used are not capable of adjustment for different levels of specification.

Application: The storey enclosure method of preliminary estimating is suitable for use during the feasibility and design phases of a construction project.

Multi-rate methods

Preliminary estimates which adopt the multi-rate method in their preparation are:

1. *Approximate quantities method:* This is generally considered to be the most reliable short-cut version of a normal method of preparing and pricing bills of quantities. Under this method, the project is measured very broadly and all-in rates are applied to their all-in quantities.

   The advantages and disadvantages of the use of the approximate quantities method may be summarised as follows:

   Advantages
   - The approximate quantities method provides the most reliable method of preliminary estimate.
   - It allows an estimate to be divided into elements or individual items if required.
   - Adjustments can be made to estimates to reflect differences in size, plan shape, storey, height and specification.

   Disadvantages
   - It can often take a considerable time to prepare the preliminary estimate.
   - Its adoption depends on the availability of reasonably detailed drawing and specification, hence it can be used only when the design is quite advanced.

   Application: The approximate quantities method is used mainly during the scheme design stage when firm design proposals are available and awaiting choice of specification.

2. *Elemental estimating method:* This method uses the elemental unit rate from elemental cost analysis of previous similar projects as a basis of the estimate. When the method is in use, the proposed project is split up into a number of elements and each element is measured (from sketch drawings) and priced using an elemental unit rate. The elemental costs are then added together to give the total estimate of the proposed project. For instance, if the foundation of a completed construction project (floor area 100 square metres) cost £1,000 then the elemental unit rate for foundation element would be £10 per square metre. Therefore, if the same type of foundation is to be adopted for the proposed project of floor area of 200 square metres, the cost of the foundation element will be

   $$\frac{£1{,}000}{100} \times 20 = £2{,}000$$

   The advantage and disadvantages of the use of the elemental estimating method may be summarised as follows:

Advantages

- It is a quick and reliable method of arriving at a realistic preliminary estimate.
- It allows a careful consideration of the cost of each element of the project.
- It is of great benefit to the designer during detail design because of the awareness of the allowance made for each element.

Disadvantages

- The method relies on the availability of reasonably detailed drawings and elemental cost analysis.
- The interpretation of differences between elements and calculating their cost effect requires a considerable skill.

Application: The elemental estimating method is used mainly during the scheme design stage when firm design proposals are available.

SUMMARY

In this chapter the initial requirements of a project at its inception has been outlined. Prime among all the requirements is the adequacy of the client's brief. Nevertheless, from time to time one reads in technical literature that the adequacy of this important tool is lacking and, hence, leads to numerous problems in the procurement of construction projects. Therefore, the client's awareness of this fact and the need for assistance in its formulation must be assessed by the client's professional advisers earlier on in the project life.

Also of great importance at the project's inception is its feasibility, viability and cost. As the preliminary estimate is the one clients remember most, the professional advisers in charge of its preparation must ensure the accuracy of this estimate. This means that the preliminary estimate must reflect all the client's project requirements.

FURTHER READING

Ashworth, A. 1988 *Cost studies of buildings*. Longman Scientific and Technical.
Cartlidge, D.P. 1976 *Construction design economics*. Hutchinson.
Ferry, D.J.; Brandon, P.S. 1994 *Cost planning of buildings*. Granada.
Gruneberg, S.; Weight, D. 1990 *Feasibility studies in construction*. Mitchell.
Piling, S.J. 1993 Brief formulation: The architect as manager. *International Journal of Architectural Management Practice and Research*, **1**, pp. 25-32.
O'Reilly, J.J.N. 1987 *Better briefing means better building*. BRE.
Rougvie, A. 1988 *Project evaluation and development*. Mitchell-CIOB.
Rutter, D.K. 1993 Construction economic; is there such a thing? *CIOB Construction Papers, No. 18.*
Seeley, I.H. 1975 *Building economics*. Macmillan.

CHAPTER 5

THE DESIGN AND ITS EVALUATION

5.1 THE DESIGN

The development of the client's brief and the design of the project run concurrently. Design is generally regarded as the principal function of the pre-construction phase and, by tradition, is the exclusive province of the architect. It commences after planning consent and approval have been obtained and the architect has investigated and surveyed the site. Although the architect normally carries out the design in stages and awaits the client's approval of each stage before proceeding to the next, he or she has the responsibility of designing a building in such a way that its tender sum does not exceed the client's development budget or cost limit. Moreover, explicit or implicit consideration needs to be given to serviceability, reliability, operability and probability of failure of all significant components in the building. The architect must ensure, at all times, that the client fully understands the design proposal and agrees that it is within his or her development budget and that they meet its objective. The design must also provide value for money in terms of both total cost and cost in use. Generally, an inadequate design which affects function, quality, aesthetics, service behaviour, initial and running costs leads to a building whose utility is often unsatisfactory and, hence, is often difficult for clients to put into use. For this reason, the architect must seek to address that issue in the design of the project.

As a rule, although the client's approval is sought in the various design stages, the design must also satisfy the architect's internal constraints (i.e. the architect's own desire to introduce additional ideas or concepts) and external constraints imposed by external sources. The architect's self-imposed constraint is the desire to explore solutions for design problems to find the one which works best technically. For example, he or she will explore design problems such as the project's priorities, shape or form of building, space requirement, arrangement of space required and the level of specification. Self-imposed constraints narrow the possibilities of solution to the design problem and thus, making the problem easier to solve. On the other hand, the architect expects to work to various external constraints imposed by the client such as technical feasibility, time and cost limit; physical factors such as site conditions and environmental standards; and external bodies such as planning control and building regulations. But, in order to arrive at a good design solution which impacts upon the satisfaction, comfort and well-being of its occupants – and if it is a commercial building, upon their productivity and performance – the architect and the design team members consider the various factors discussed below.

5.2 DESIGN EVALUATION

As the initial budget estimate is always the one that is remembered by a client, steps must be taken to ensure that the initial figure is not exceeded. Normally, the design of the building influences its eventual overall cost, therefore, over the years, increasing pressure has been exerted on the design team to evaluate the cost, quality and design before the project leaves the drawing board. This evaluation usually involves an examination by the design team of all viable design options at various stages in the design sequence and maintenance of cost control by undertaking a series of cost checks. The cost checks are a series of on-going exercises, carried out during and in conjunction with the design process in order that the cost effect of design decisions can be reported and examined. If the trend of cost checks indicates an overall increase in the project cost, corrective action is taken.

5.2.1 Objectives of design evaluation

A design evaluation exercise is undertaken on construction projects for the attainment of a multitude of objectives, including the following:

- To ensure that the sum of money which the client sets out to spend on the construction project is not exceeded.
- To control the cost of a project during the design phase so that the eventual tender figure can be predicted with a degree of certainty.
- To provide a balanced design expenditure throughout the construction project, thus giving the client value for money in terms of a functional, well-designed and soundly constructed building.
- To provide the basis of comparing cost implications of different construction projects.

The above objectives are of importance to the client. If, for instance, the client development budget is exceeded, it may be found rather difficult to raise the additional finance needed to complete the project. Moreover, when placed in an additional borrowing situation, the client's expected margin of profit in the development might shrink to almost nothing and all investment efforts would then be wasted.

5.2.2 Need for design evaluation

In addition to the need to ensure that the client's budget is not exceeded, the following economic considerations have created the need to check the construction project cost constantly during the design phase:

- Clients have become more exacting and cost conscious for reasons of profitability and accountability. As a result, construction costs are being scrutinised more closely and with greater skill and accuracy.
- Construction projects are now larger, are constantly of technical complexity and, hence, are more expensive.
- There is an increasing sophistication in funding arrangements for construction projects requiring a more efficient usage of funds, and higher accountability.
- Modern construction projects are often exposed to rising prices, restrictions on the use of capital, high financing costs and a volatile financial market.
- There is a greater awareness of life cycle costs as a result of utilisation of new materials, components and methods of construction.
- There is now greater emphasis on a faster pace of the construction process. This creates the need for the development of an optimum design as there is no time to redesign if the lowest tender was excessively high.
- Rising costs of construction resources has introduced a general trend towards greater cost effectiveness and a move to reduce or eliminate waste where possible.
- Clients have multi and complex requirements which require study and assessment of their full financial implications before incorporation into a construction project.

If the design evaluation process was postponed until the end of the design stage, serious problems such as redesigning and consequent delays might arise. But by undertaking the evaluation process as design progresses, the design team are able to cost-design within the client's budget. In addition, the design team are able to obtain information with regard to initial and future costs so that design decisions are made with full knowledge of the financial implications of that decision.

5.2.3 Design evaluation process

The design evaluation process is composed of distinct but interrelated functions of cost planning, life-cycle costing and cost studies (see Figure 5.1)

5.3 COST PLANNING

The process of cost distribution, monitoring and control during the project design stage is generally known as *cost planning*. The purpose of cost planning is:

- Provision of cost information on the proposed total project expenditure to assist the design team members in design decision making.

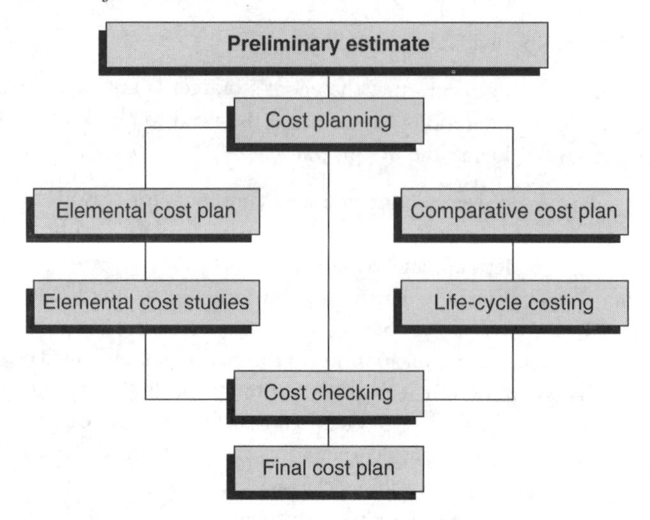

Fig. 5.1 Composition of pre-contract cost management

- Rational distribution of the budget to avoid large sums of money being spent on one element to the detriment of others.
- To facilitate the study of the economics of the relationship between initial capital cost and maintenance costs of materials/components to be used on the construction project.

The above objectives are obtained by checking the cost of each element of a building as the design progresses. It is worth mentioning at this point that cost planning is a subject of depth and intricacy. Hence, it must be emphasised that the treatment in this volume is purely by way of summary and should be regarded as a series of signposts pointing to the field of further study.

5.3.1 Cost planning techniques

Preparation of a cost plan involves critical breakdown of the preliminary estimate or the total funds to be spent on a construction project (i.e. the client's cost limit) and allocating the constituent amount to the various elements for the proposed building (see Example 5.1). As the design progresses, checks are made against design and the necessary financial adjustments made among the elements. By so doing, the probable cost of the project will thus be known within close limits at all times while various design decisions are being made and, therefore, it is essential that the cost plan is understood by all the parties involved in the pre-contract phase of a construction project.

In the design decision-making process, a complete system of cost planning would comprise elemental cost plan, comparative cost plan, cost checking and life-cycle costing.

Example 5.1 Presentation of cost plan

UPPER DAGNALL STREET DEVELOPMENT

COST PLAN NUMBER 1 **Date:**

Elements	Outline specification	Total cost of element (£)	Cost per m² gross floor area (£)	Element unit quantity (m²)	Element unit rate (£)
1. Substructure	In-situ concrete bored piles average 25 m long (approx. 120 No) reinforced concrete pile caps and ground beam, 125 mm thick floor planks	180,000.00	72.00	854	215.83
2A. Frame	Reinforced concrete beams and columns.	270,000.00	108.00	2,500	108.00
2B. Upper Floors	150 mm reinforced in-situ concrete slab (all floors).	78,000.00	31.20	1,668	46.76
2C. Roof	125 mm Reinforced in-situ concrete; insulated; 25 mm screed and 25 mm mastic asphalt.	30,000.00	12.00	834	35.97
2D. Stairs	1200 mm wide reinforced in-situ concrete with mild steel balustrading.	6,000.00	2.40	-	-
2E. External Walls	280 mm brick hollow wall; outer skin. Handmade Sandringham P.C. £300.00/thousand; inner skin 100 mm concrete blockwork	102,000.00	40.80	1,285	79.38
REMAINDER OF ELEMENTS		1,734,000.00	693.60		
		2,400,000.00	960.00		
	Preliminaries (say) 10%	240,000.00	96.00		
	Contingencies (say) 2.5%	60,000.00	24.00		
	Design risk allowance (say)	50,000.00	20.00		
		£2,750,000.00	£1,100.00		

Elemental cost planning

In elemental cost planning, the designer's design solution is initially stated in sums of money which represent the design budget. The designer then proceeds to develop the design within the framework of what he or she believes to be the correct economic approach, but not exceeding the design budget. For this reason, elemental cost planning can be referred to as 'designing to cost'. In the process, the quantity surveyor assists the designer in the choice of correct economic framework, provides cost information and checks the designer's design solutions against the predetermined project budget. Additionally, it also means that the designer may have to redesign any solution that has a cost that does not fit in with the cost plan.

Preparation of an elemental cost plan Assuming the client has agreed the cost limit during the briefing stage and the quantity surveyor has provided an agreed outline cost plan at the beginning of the design phase, the design team will want to go ahead with the preparation of working drawings. The quantity surveyor will commence the preparation of the cost plan based on the following information:

- Copies of sketch plans including drawings showing elevations of the proposed building.
- A brief specification of the proposed method of construction.
- Information on proposed standard of finishings, fittings and so on.
- Proposed tender date and contract period.
- Probable mode of contractor selection.
- A suitable cost analysis.

The steps generally adopted by the quantity surveyor in the preparation of an elemental cost plan are as follows:

1. The total floor area of the building is calculated from the designer's sketch plans.
2. Selection is made of a suitable elemental cost analysis of a similar project with which the quantity surveyor is familiar. Where this is not available, an elemental cost analysis of a project similar to the proposed building in terms of size, plan shape, total height and specification is selected from a published list of elemental cost analysis.
3. The chosen elemental cost analysis is adjusted for price changes, variation in type of contract, special local considerations and so on (see preliminary estimate of costs, Example 4.1, page 66).
4. The total areas of the various elements measured from the sketch plans are multiplied by the elemental cost per square metre floor area to obtain the total cost of an element. This figure is compared to that in the outline cost plan for action (i.e. affirmation or redesign of an element under consideration).

Comparative cost planning

As in the case of the elemental cost plan, it is assumed that some discussion of costs and cost limit for the proposed construction project has been agreed during the briefing stage. But unlike the elemental cost plan, the comparative method does not seek to enforce rigid cost limits for the design of particular elements but more to maintain the flexibility of choice of a combination of possible design solutions. It is more concerned with the comparison of alternative possibilities within the total sum allocated to an element. It is generally recognised that if the most economic scheme is to be designed, all the alternative design solutions should be studied to determine their cost implications. Hence, during the design phase, the designer will have to explore a range of alternative solutions and select an optimum initial and final solution that is satisfactory to the client's requirements. It is therefore, the responsibility of the quantity surveyor to provide the other design team members with the cost consequences of alternative design solutions to enable the architect to make a selection and develop the design.

Comparative cost planning methods have the advantage of comparatively attempting to tackle a systematic search for a possible optimum solution which is appropriate to the target cost. However, it is extremely time consuming and is also difficult to produce a satisfactory range of descriptions that will be simple enough for the design team members to incorporate into the design concept.

Cost in use

In the preparation of comparative cost planning, user cost is one of the components of construction costs that may be considered. In considering the user costs in addition to the capital costs, the quantity surveyor is in effect considering the total construction costs made up of capital, operating and maintenance costs. However, all these costs (capital, operating and maintenance) occur at different points in time and they must be brought to a common time base for comparison. For this reason, in order to compare the future payments with the initial expenditure in cost in use calculations, it is necessary to convert the future expenditure to their present discounted value. They can therefore be discounted and the results expressed either in terms of net present value (NPV) or the equivalent annual value. From this exercise, the most economic material or component is selected in line with the client's development strategy.

Cost checking

As the project design develops, it becomes necessary for the quantity surveyor to carry out a process of checking the estimated cost of each element against the cost target in the agreed cost plan. These checks are necessary if the client's cost limit is not to be exceeded. Cost checks on working drawings can be a time-consuming exercise, especially where the design has been altered or changed for a particular reason. Where the design alterations are drastic, it may be quicker for the quantity surveyor to prepare a new cost plan rather than attempt to check the revised drawing against the cost plan. However, the extent of cost checking carried out on any project will depend upon the time available, the extent of alteration from the cost plan, the familiarity of the project, the degree of confidence between the design team and the cost sensitivity of the element.

Generally, during the cost-checking process, if the cost of an element differs from the cost plan, it will be necessary to adjust the cost of other elements that make a balance or perhaps draw on reserves set aside in the cost plan. However, if it proves impossible to keep within the set target, the quantity surveyor must report this to the designer who may consider a different design solution or accept the increased cost of the element and increased cost of the project as well.

Elemental cost studies

Elemental cost studies are undertaken during the earliest stages of the project to aid the selection of an optimum design solution. The study involves a breakdown of the total building cost into elemental costs for analysis to determine how these cost correctly reflect a given situation or to relate the cost of an element to its function in a construction project. Current economic trends, technological advancement and/or change, new construction techniques and optimum conditions have an impact on the efficiency of design solutions and the quantity surveyor must ensure that they are correctly reviewed and represented in the elemental cost studies.

Also, represented in the elemental cost studies is the establishment of how alternative design solutions can be evaluated to obtain the optimum solution. For example, in the selection of a frame for multi-storey structure, the optimum solution is likely to be either steel or reinforced concrete. Here, the elemental cost study will look into cost items such as cost of materials; frame design; speed of design and erection; availability of materials, plant and skilled operatives; location of project; fire-resisting properties; weight of material; effect on foundation design and so on. The effect of all the above factors on the project should be studied in detail and the optimum design solution selected to suit the site conditions and client's development requirements. It must be remembered also that elemental cost studies are a time-consuming exercise and should therefore be undertaken for the more cost-sensitive elements of a project.

5.3.2 Final cost plan

After the quantity surveyor has carried out all the cost checks and has effected all the necessary financial adjustments, it is then time to prepare the final cost plan. The final cost plan should be presented as neatly as possible and outline the agreed specification on which the cost plan was based. A copy of the final cost plan should go to the client and each of the design team members for future reference (see Example 5.1).

5.4 DESIGN FACTORS AFFECTING BUILDING COST

In order to design a building within the client's cost limit, there is a need to adopt an economic approach to design. For this reason, during the cost planning process the design team members will consider the relationship between a change in design and its corresponding change in cost. In the process, the following factors, which have a direct influence on the cost of a construction project, are considered.

1. *Size of building:* Generally a client's functional requirements determine the size of a building. Although the size usually dictates construction costs, selling price or rental income, construction costs are not proportional to changes in building

size. Moreover, the increases in the size of buildings usually produce reductions in unit cost per square metre of floor area. Nevertheless, in all construction projects, the interrelationship of size, construction costs and revenue are the important factors which influence the construction project's design decisions. Normally for each construction project, the above variables are different and, hence, the design team devise an optimum design solution that gives the client a minimum cost for a maximum revenue. For this reason, the size of the client's building should be increased only if the additional margin of profit is sufficient to warrant the additional expenditure.

2. *Plan shape:* The plan shape of a building contributes to its pleasing appearance. While the designer aims at the achievement of a pleasant building, it must be remembered that the plan shape of a building has an influence on the overall cost of the construction project. Plan shape is often dictated by the shape of a site; however, a change in the plan shape of a building, even though other components remain unchanged, affects the perimeter costs. The perimeter costs of a building include external walls, windows, external doors, foundations, parapets, perimeter columns and so on. Therefore, the more a perimeter wall can be reduced without reducing the floor area the more economical a scheme is likely to be. From the above analogy, it can be concluded that a square plan shape building provides the lowest amount of wall area compared to gross floor area (known as wall to floor ratio). The lower this ratio, the more optimum the plan shape. Example 5.2 demonstrates the wall to floor ratio for a single-storey building, 3 metres high, and from this demonstration building plan shape *A* is the most economical. Building shape also influences the cost of the following:

Example 5.2 Buildings – walls to floor ratio

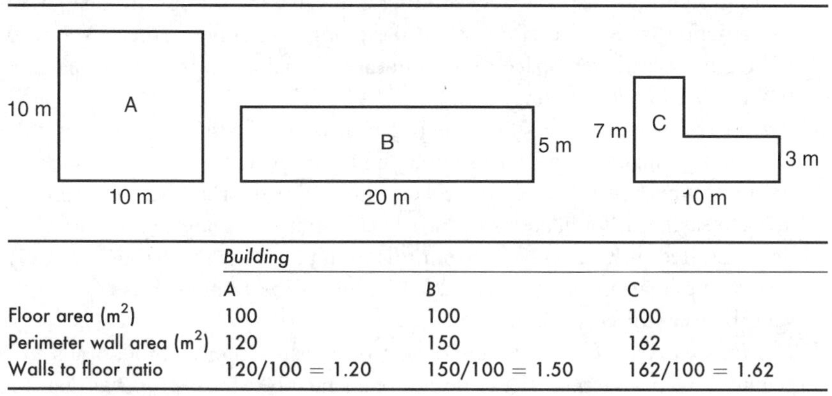

	Building		
	A	B	C
Floor area (m^2)	100	100	100
Perimeter wall area (m^2)	120	150	162
Walls to floor ratio	120/100 = 1.20	150/100 = 1.50	162/100 = 1.62

- Internal partition costs – the larger the girth of perimeter wall, the higher the cost of partitioning.
- Lighting, heating, mechanical and air conditioning costs – the larger the girth of the perimeter wall, the higher the rate of heat loses and/or gain.

- Setting out, drainage and external works costs – the larger the girth of the perimeter wall, the longer the drain runs, the number of manholes and connections increase and external works also increase.

The design team members therefore consider all the above factors when deciding on the optimum economic shape.

3. *Total height of building:* As a general rule, the cost per square metre increases with the number of storeys. This cost arises as a result of a number of factors, such as extra loading on the foundations, provision of means of access (staircases and lifts), loss of productive time (operatives taking longer time to reach workplace at higher levels), improvement in fire resistance and so forth. The design team therefore normally consider the cost and benefits of placing some storeys below ground level.
4. *Storey height:* The storey height of a building is determined by its intended use and is subject to the approval of the local planning authorities. Generally, variations in storey heights cause changes in the cost of the building. The main cost items affected by increasing the storey heights of a building are both external/internal walls and partitions, finishings and decorations. Additionally, due to the increased volume of the building, additional heating may be necessary as well as some extra lengths of service pipes and cables. Therefore, the design team normally consider the above costs and many others when making optimum design solutions on storey heights of buildings.
5. *Communication/circulation space:* The amount of circulation space provided in a building affects the economy of its production cost. Circulation space comprises entrance halls, passage, corridors, access decks, stairways and lift wells. These are regarded as dead space yet they involve considerable cost in heating, lighting, cleaning, redecorating and rates in addition to their original construction costs. Therefore, when designing for economy there is a need to reduce the circulation space to a minimum compatible with the satisfactory functioning of the building.
6. *Arrangement of buildings:* When the design team members are considering the optimum arrangement of, for example, housing and industrial units, interlinking the units where possible leads to cost savings. The interlinking arrangements leads to saving in foundations, external walls, finishings and decorating as demonstrated in Example 5.3. From this example it can be seen that there is a cost saving in arrangement Type 2 as a result of the combined use of separating walls between three structures.
7. *Construction method:* The selection of good construction methods leads to economy of production. If the design team members take proper account of the production process, they can reduce inefficiency in the use of resources such as idle time and discontinuity in the work of particular trades. For example, the use of prefabricated components counteract high costs of site labour and supervision. Additionally, the economies of the use of steel frame as opposed to reinforced in-situ concrete are normally considered for optimum design solution.

Example 5.3 Building arrangements

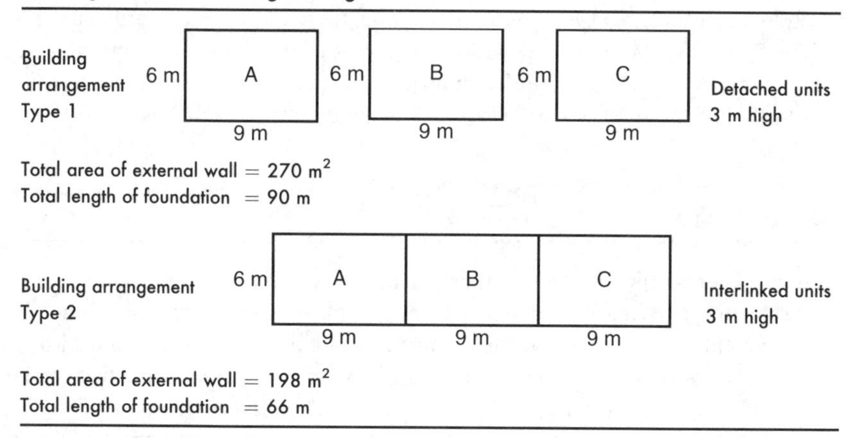

5.5 BUILDABILITY

Designing for buildability implies the elimination of non-productive site processes, ease of production and increased efficiency of site management. Generally, there are a number of items in design that can affect the flow of site production and, therefore, can also affect the efficient use of time and resources. Hence, the design team members adopt the following design measures in order to achieve buildability of a construction project.

- Simplification of design and the creation of standard or relatively few technically different activities to promote repetitive production cycles. For example, column and beam dimensions may be standardised to allow repetitive use of formwork.
- Design dimensionally coordinated to reduce the labour of on-site cutting of materials to waste. Additionally, workable/sensible dimensions and tolerances given on working drawings.
- Design details kept simple and adhere to the tried and tested. Moreover, unique and one-off features/details avoided.
- Design employing readily replaceable resources. For example, the materials or components selected should be traditional and readily available in the locality.
- Selection of materials or components and the method of their incorporation into the building should be the type that has been tried, tested and well understood by all the project's participants.
- The construction method selected are tailored to suit the site or locality and to make the best use of the available skill and resources.
- Services designed for their efficiency in operation and ease of installation. Design coordination of items such as service ducts, spaces for machinery and equipment, wall chases for pipes, conduits and lighting fixtures can save time and money initially and on maintenance costs ultimately.
- Prior arrangements made with statutory authorities for supplies to be made

available by the time the construction project is ready for occupation. Such advanced arrangement is essential as it facilitates planning and coordinated layout. It also avoids costly delays and conflicts of the building's physical space requirements, its external works, drainage and underground distribution mains with that of the statutory authorities.

5.6 QUALITY ISSUES

Designing for quality implies a design solution which meets the client's requirements in terms of function, aesthetics and external attributes. The establishment of a benchmark for quality is difficult even for a client with unlimited resources, and is even more difficult to measure and control during design. However, it is generally accepted that, while the assessment of quality of construction is a subjective matter, quality can be measured against design drawings and specification. Therefore, when designing for quality, the design team members need to integrate all quality-related efforts economically to satisfy the client's stated or implied needs. In this regard, the design is rationalised to a simplified construction approach and, additionally, quality levels are clearly specified in the contract documentation. Although quality of design is ambiguous and a matter of individual judgement, the design team members pay particular regard to the following in order to meet the client's required quality.

- Client's briefings are becoming more sophisticated, and therefore, the design team members provide clients with all the assistance required in the formulation of a brief and the establishment of quality levels for the development.
- Establishment of client needs and wants and, moreover, giving priority to needs.
- Planning the project's design adequately and also checking the accuracy and reliability of design information.
- Giving due appreciation to the practicality of the project's design, efficient production techniques and familiarity with skills and materials available in the project's locality.
- Removal of uncertainties of the client's initial requirements from the design team members by the establishment of good relationship and good communication medium for free flow of design information.
- Selection of tried and tested material/components and providing information concerning quality continuously throughout the project.
- Recognition that drawing is a means of communication rather than an end product and, hence, should be carefully detailed and coordinated to provide good production information.
- Appreciation of the fact that clear and adequate information on the plan shape, size and location; constituent parts of building; adequate specification on materials; jointing and fixing methods conveys the designer's intentions adequately to the builder.

- Thinking ahead, carrying out project design reviews or envisaging possible construction problems that can be met with on site and provide solution at design phase.
- Provision of value for money not through a cheap design solution, but by understanding the client's quality requirements; selection and balance of cost; construction projects function, time and appearance.
- Awareness of environmental issues and incorporating energy-saving measures into the project's design so as to achieve energy efficiency.

5.7 EXPEDITED PRODUCTION

In construction projects where time is of the essence, the design team members try during the design phase to identify and provide solutions for production problems likely to arise on site and thereby facilitate the adoption of expedited production. In addition, the design team members usually adopt the following measures to achieve the objective of fast production:

- Adequate site and/or soil investigation prior to commencement of the construction project design.
- Design details kept simple; unique and one-off features avoided.
- Column, beam and wall dimensions standardised to allow repetitive use of formwork.
- The materials and components selected and the method of their incorporation into the project are such as will be easily understood by all project participants.
- The selected materials and components are those readily available in the project's locality; advance purchasing of same where necessary.
- Steel frame construction and composite metal decking or precast slab floor instead of in-situ reinforced concrete.
- Specification for high early-strength concrete instead of ordinary Portland cement.
- Use of precast concrete stair flights instead of casting them in-situ.
- Avoidance of wet trades on the critical path and replacing, for example, wet steel casing/fire protection with dry lining.
- Off-site prefabrication reduces the number of workmen on site and also minimises wet trades (e.g. prefabricated bolt-on toilets, caged column and beam reinforcement and composite steel trusses designed to be prefabricated and dropped into position by tower crane directly from delivery lorries).
- Items of heavy plant (e.g. boiler, refrigerator plant, main air-handling plant, lift motors, sprinkler pumps, sub-station) located in the basement to enable early start on installation to be made; only small and packaged plant items are located on the roof (e.g. water tank, lift pulley room, cooling tower).
- Downstand beams and drop-ends around columns eliminated to allow for decking out and flush soffit if in-situ concrete construction cannot be avoided.
- Precast or prefabricated cladding material chosen to provide an early weatherproof envelope.

- Pre-assembled mechanical/electrical units and pipework sections with an acoustic and thermal insulation cover averts the need to build a heavily insulated plant room.

5.8 SAFETY ISSUES

In designing for safety, the design team members aim to provide a design solution which is not only safe to the environment, but is also safe to produce and enjoy. In the process, the design team members carry out forward planning to identify potential safety problems at the design phase and make provisions for their avoidance. As there is enough evidence to suggest that a greater proportion of occupational accidents during the production phase are inadvertently pre-programmed before commencement of production, the design team members normally adopt the following measures during the design phase to achieve safety objectives:

- A safety adviser or coordinator is appointed to help the establishment of safety objectives in design and their regular evaluation.
- Positive account is taken of health and safety hazards as a design consideration among other considerations.
- Applying, when relevant, the principles of prevention, reduction or control of hazard during production.
- The design lending itself to safe means of production by thorough and rational planning at the design phase and specification for safe production methods.
- Informing the authorities responsible for health and safety in advance of the construction project of any dangerous works associated with it.
- Providing any known information on production hazards so that the prospective contractors can plan for safety and the cost implications.
- Asking for safety policy and assess safety competence of prospective competing contractors.
- Study of the likelihood of the construction project producing a particularly complex or adverse effect on the environment, such as discharge of pollutants (e.g. water, air, soil, noise, vibration, light, heat and radiation) and seek advice on ways and means of modifying or mitigating its effects.
- Desisting from specifying materials and components harmful to the environment, operatives and the ultimate users of the completed facility (e.g. rigid urethane foams, extruded expanded polystyrene foams, phenolic foams, urea formaldehyde foam cavity insulation, timber and timber products that are not from managed and regulated sustainable sources, lead-based paints and asbestos including chrysotile).

5.9 TAX EFFICIENCY

In the UK, certain types of construction products, such as offices and factory buildings, are eligible for capital allowances. The design team members are therefore made aware of the impact that tax may have on certain design decisions, and are thereby able to produce tax efficient designs. By considering tax efficiency of building design, the design team members assist the client to realise full entitlement to tax relief on a construction project. Therefore, where appropriate, a design audit is carried out on the design when fully completed. The purpose of the audit is, first to establish and confirm the client's investment requirement and to ensure that these have been interpreted accurately in the design; and, second, the design audit is to establish the legal basis upon which to make a capital allowance claim.

Capital allowances are many and varied. In the UK, they are usually offered on a sector basis, regional basis or both and the most common types of allowances on property may be listed as follows:

- Plant and machinery allowances.
- Industrial building allowances.
- Hotel allowances.
- Scientific research allowances.
- Enterprise allowances (as when development takes place in an enterprise zone or area of recognised high unemployment).

Designing for tax efficiency demands forward planning which involves the identification of all potential allowances at an early stage in the design phase, and modifying and forecasting the client's legal entitlement. The possession of this information enables the client to negotiate with the Inland Revenue or Income Tax authorities for tax relief at the design phase. These tax incentives are also taken into account when preparing budgets for the type of development concerned, and when a development qualifies for capital allowances these are allowed as follows:

1. *Initial allowance:* This is an initial tax allowance made to a client for incurring capital expenditure on the construction of a new building or structure.
2. *Writing-down allowance:* This writing-down allowance is a year by year sum allowed by the Inland Revenue to the client for a specified period.
3. *Balancing allowance or charges:* Balancing allowance and charges is given to a client when a building structure is sold for a sum lower than its written-down value. However, the converse is that the Inland Revenue makes a balancing charge against the client when the same building is sold in future for a sum higher than the written-down value.

5.10 ENERGY EFFICIENCY

As public awareness and opinion is pointing towards conservation of natural resources and, moreover, as the price of energy rises, it is beneficial to the client if his or her building is designed to save energy. What this means to the client is a considerable reduction of energy wastage during the use of the building and, hence, payment of less gas and/or electricity bills. For this reason, the client's design team members give a careful account of design which will reduce energy consumption. Of particular importance are design considerations for:

1. *Orientation and degree of exposure:* Buildings on exposed and elevated sites do not contribute to energy economy. Buildings on exposed sites are affected by increased wind speed which increases ventilation rates and hence, decrease surface resistance. Elevated sites are generally colder and, moreover, are subject to more mist or rain and therefore buildings on such locations do consume more energy.
2. *Layout and grouping of buildings:* The layout of buildings should aim to minimise wind effects and allow maximum solar penetration. Also, grouping of individual buildings into composite blocks has a major effect on energy conservation.
3. *Plan, form and shape:* The squarer the plan for a given floor area, the less the energy wastage. The square contains a maximum area related to length of the perimeter (see Example 5.4). However, the reverse is true in lower blocks as the surface area is exposed to increased wind speed and, hence, increased ventilation.

Example 5.4 Plan area related to girth of building

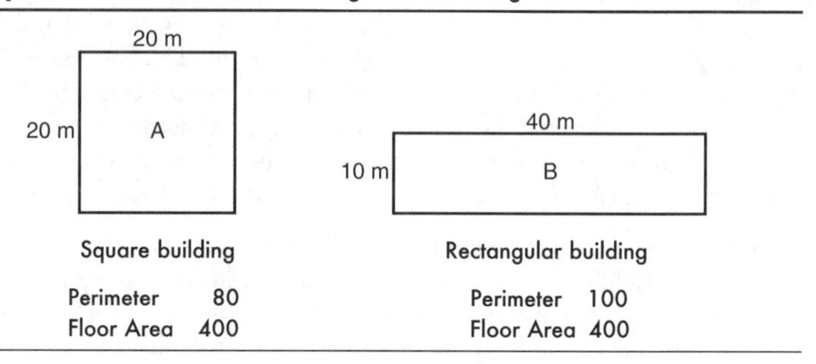

4. *Interior planning:* Open plans do not lend themselves to energy conservation since no partial use or control of energy is possible. Also, where feasible, heat source such as boiler and flue should be given internal location for efficient performance and also to contribute to heating the upper floors.
5. *Amount of ventilation:* Reduction of ventilation or air change rates conserves energy. However, a balance should be struck between the reduced ventilation level and the risk of condensation.
6. *Thermal installation:* The system of heating selected should have both thermostatic and time controls to respond to changes in heating requirements.
7. *Thermal performance of the fabric:* Materials and components used to envelop a building should be able to provide adequate insulation and appropriate thermal response. Thermal performance involves a measure of heat transfer, the outside influences affecting thermal capacity, surface temperature, conductivity, solar absorption and air infiltration. All these require counter-action of the energy input into the building itself to provide thermal comforts.

For this reason, the client's design team must pay particular regard to the above influences so that they can select materials/components to suit the relationship between exposure and internal activity, and the masking envelope and energy input. Also, in order to produce an economic solution, there is the need for an inclusion of a specialist energy management engineer in the design team to advise on valid energy conservation measures, taking into account the type of heating, solar and other energy gains and assessing the effects of design decisions on energy efficiency. There is also the need for a life-cycle costing study on all design alternatives considered so that a suitable solution, providing an optimum balance over as long a period of the annual cycle as possible, is arrived at for the client's benefit.

SUMMARY

The success of the efficient design processes described in this chapter depends on corporative efforts of the client and all the design team members. This is achieved by several meetings and discussions in which the client provides the design team members with information needed for greater appreciation of the brief and its technical interpretation and development to suit the client's budget. The design solution is therefore one of a range of possible solutions constrained by the architect's perceptions of the client's evolving briefing process. Therefore, the efficient design control processes can be said to be evolving also and characterised as a series of interactions between all participants, with solutions being tested for effectiveness at each stage of development. The process continues until the final solution satisfies the client's technical, safety, aesthetics, quality, function and cost requirements.

Cost planning is a generic term describing the various methods adopted to shape the construction project's budget as design progresses. The method adopted depends on the cost information required; however, on large projects all the methods outlined above are employed for an effective pre-contract cost management.

FURTHER READING

Ashworth, A. 1988 *Cost studies of buildings*. Longman Scientific and Technical.

Barthurst, P.E.; Butler, D.A. 1973 *Building cost control techniques and economics*. Heinemann.

Bloomfield, J. 1993 Value added quality. *Chartered Builder*, April, pp. 12–13.

BRECSU 1993 Energy efficiency in new housing. *Good Practice Guide No. 79*, Building Research Establishment.

Cartlidge, D.P. 1973 *Cost planning and building economics*. Hutchinson.

Cartlidge, D.P. 1976 *Construction design economics*. Hutchinson.

Chevin, D. 1993 Design counsel. *Building*, November, pp. 36–7.

Coles, E. 1990 Design manager: need training. *Chartered Builder*, December, pp. 16–17.

Ferry, D.J.; Brandon, P.S. 1994 *Cost planning of buildings.* Granada.
Harper, D.R. 1990 *Building – the process and the product.* The Chartered Institute of Building.
Lai, H. 1989 Integrating total quality and buildability; a model for success in construction. *CIOB Technical Information Service No. 109.*
Llewellyn, T. 1993 Why architects should plan for a tax-efficient design. *The Architects Journal*, November pp. 28–9.
Seeley, I.H. 1975 *Building economics.* Macmillan.

CHAPTER 6

FORMATION OF A VALID CONTRACT

6.1 INTRODUCTION

Due to the intricate nature of the construction process, there is a need for a valid agreement prior to commencement of production. But before discussing the various standard forms of building contract currently used in drawing up an agreement, it is useful to survey in outline some of the legal principles that govern formal binding agreement. It must be stated, however, that the law of contract is profound and intricate in nature and therefore its full treatment is outside the scope of this volume. For this reason, it must be emphasised that the coverage of this volume is purely a summary and, as such, should be regarded as a pointer to the domain of legal reference books and established legal opinion.

6.2 DEFINITION OF A CONTRACT

A contract is a legally binding agreement between two or more persons. Legally binding means that the agreement will be enforced by the Courts. The breach of an agreement that is legally binding allows the injured party to claim a legally enforceable remedy from the party who commits the breach.

6.3 ESSENTIALS OF A VALID CONTRACT

Not all agreements constitute legally binding contracts. For a valid contract to come into existence, certain essential requirements must be met with respect to the nature and circumstances which surround the reaching and making of that agreement. The agreement must contain the following features, which are essential in order to enter into a valid contract.

6.3.1 Offer

An offer is a proposal by one party of its willingness to be legally bound by the proposal as soon as it is accepted by another party. An offer may be made in writing, in words or by conduct and terminates if:

- It is rejected by the party to whom the offer was made.
- It is revoked by the offeror (a party making the proposal) before acceptance.
- Either party dies before acceptance.
- A stipulated time limit expires before acceptance.
- It has not been accepted within a reasonable time (i.e. where no time limit has been specified for acceptance).

6.3.2 Acceptance

An acceptance is a full acceptance by a party to whom an offer has been made. However, a valid acceptance requires an unconditional assent by the offeree (the party to whom the offer is made) to all the terms of the offer. Furthermore, the acceptance must be communicated to the offeror and this may be made in writing, in words or by conduct.

6.3.3 Intention to be legally bound

It is necessary that the parties can be deemed to have intended to create legal relationship. This intention makes the agreement they have reached to be one which would be legally enforceable.

6.3.4 Capacity to contract

Capacity refers to the legal status of parties to make a legally binding contract. Generally, all persons have full and unqualified status to be bound and to bind others by agreement. The exceptions to this general rule may be summarised under the following:

- Minors – i.e. persons under the age of eighteen – will not be bound by certain types of agreement. However, as a general rule, a minor may enter into a legally binding contract for his or her benefit and necessities.
- Corporations are artificial persons and, hence, the capacity of corporations to make legally binding contracts depends on how the corporation was created.
- Drunkards and persons of unsound mind.

6.3.5 Consideration

Consideration is something of legal value given, done or forborne by one party in return for an action or inaction on the part of another party. The general rules which validate consideration are:

- Consideration must be real, but need not be adequate and must move from the promisee (i.e. parties intending to enter into contract must provide the consideration).
- Consideration must not be vague and must not be past.
- Consideration must be legal and something beyond the promisee's existing obligations to the promisor.
- Consideration must be possible (i.e. it must be capable of fulfilment at the time the contract is made).

6.3.6 Genuine consent

It is essential that the agreement is made with proper and genuine consent by the parties to it. Therefore, a party's consent must not be induced or caused by such circumstances as false statements, fundamental mistakes, misapprehension and undue pressure exerted by the other party.

6.3.7 Legality

The agreement must not involve an illegal purpose or activity. For instance, an agreement to defraud the Inland Revenue is not legally enforceable.

6.4 CONTRACT TERMS

The terms of a contract are provisions or stipulations in a valid contract describing some aspects of the agreement between the parties to contract. They define the rights and obligations of the parties to each other and the extent to which they are in agreement. The terms may be express, implied or statutory and are prevalent in all construction contacts.

6.4.1 Express terms

Express terms are the terms expressly agreed by the parties to contract and by which they intend to be bound. They are words expressed orally or recorded in handwriting, typing or printing by the parties. In most construction contracts, express terms usually take the following forms:

1. *The agreement:* This is the written details of the project and the agreed sum payable for its completion.
2. *The conditions of contract:* These are the detailed provisions governing the execution and administration of the project. Usually, there will be a standard form of conditions drawn up and published by various representative bodies (see section 6.1.0).
3. *Drawings:* This is a document delineating the plan shape of the project as well as its design details.
4. *Specification or bills of quantities:* A document describing the quality of the materials/workmanship as well as the quantity of works required.

6.4.2 Implied terms

Implied terms are contract terms which are not written down in a contract or openly expressed at the time the contract is made, but which the law implies. For instance, in the absence of express terms, the following term will be implied in a construction contract.

> *That the building contractor will execute the works in a workmanlike manner and complete within a reasonable period of time.*

6.4.3 Statutory terms

Statutory terms are contract terms which are imparted into contracts by legislation. An example of this in a construction contract is The Supply of Goods and Services Act 1982.

6.5 UNENFORCEABLE CONTRACTS

Contracts that become unenforceable may be described as void or voidable contracts. A contract which is void does not give rise to legal rights and obligations and, hence, cannot be sued upon.

6.5.1 Void contracts

A contract may be void in any of the following circumstances:

Mistake

A mistake in a contract may arise when:

- Agreement has been reached but upon the basis of a common mistake (e.g. where the mistake relates to the existence of a subject matter which is non-existent at the time of contract). In this situation, any contract entered into is void by reason of non-existence of subject matter of the contract.
- Parties have different intentions. In this situation, when the Court cannot ascertain the true meaning of the contract, then the contract is void.
- Parties have negotiated at cross purposes.
- One party has made a mistake and the other party is aware of the mistake.

When the situation in the last two terms occurs, the acceptance may not correspond with the offer and the validity of the agreement may be in doubt and, hence, voidable.

Misrepresentation

Misrepresentation may be described as the making of an untrue statement relating to fact which induces another party to enter into contract.

Misrepresentation may be fraudulent, innocent or negligent and the general effect is to render the contract voidable by the injured party who might have relied on the untrue statement. The misrepresentation must also have been a material cause of his or her entering into the contract. Misrepresentation may be categorised as follows:

- Fraudulent misrepresentation comprises an untrue statement made recklessly with full knowledge that it is untrue.
- Innocent misrepresentation is an untrue statement that is made in the belief that it is true.
- Negligent misrepresentation is an untrue statement that is recklessly made in the belief that it is true (i.e. without checking the authenticity of the statement).

When misrepresentation occurs, the injured party can either affirm or repudiate the contract. He or she can also bring an action for either recession and restitution or damages.

Illegal contracts

Contracts which contain an illegal or wrongful element are in general void and no action may be brought upon them. Examples of illegal contracts are contracts prohibited by statutes; contracts to defraud the Inland Revenue; contracts against the interest of the state, contracts leading to corruption in public life; and contracts which interfere with the course of justice. Such contracts cannot be enforced by either party.

6.6 PRIVITY OF CONTRACT

Under the doctrine of privity of contract, no party may be party to or bound by the terms of a contract to which he or she is not an original party. For this reason, a contract can neither impose obligations nor confer rights upon others who are not privy to it. However, there are exceptions to this rule as in the case of actions by agents and transactions involving properties.

6.7 AGENCY

Agency is a general term which describes the relationship between two parties whereby one, the agent, acts as the representative of the other, the principal. Therefore, under this relationship, an agent is invested with a legal authority to establish contractual relationships between his or her principal and a third party. Examples of agents in the construction industry are architects and engineers and, therefore, providing an architect or an engineer acts within the confines of the legal authority conferred on him or her by the principal, his or her acts become those of the principal. Therefore, legally, the principal (client) must accept the responsibilities for the decisions made by the architect or the engineer (agents) in the course of performing their duties on a construction project.

6.8 PARTIES TO A CONSTRUCTION CONTRACT

As noted above, any valid contract has two parties, the offeror and the offeree. For this reason, there are only two parties to a construction contract – the client (also known as the employer) and the building contractor. But owing to their complex nature, construction projects usually consist of an intricate network of different contracts between several parties (see Figure 6.1). The contract between the client and the building contractor is known as the main contract. This contract covers the agreement between the two parties only and hence the other parties shown on Figure 6.1 are not parties to the main contract and derive no obligations or rights from it. They are nevertheless parties to other contracts within the overall structure of the construction process, and derive their rights and obligations under other contracts to which they are parties. However, owing to the customary divisions of duties and responsibilities within the construction process, some of these persons are named in the main contract.

The basis of the contract is the offer by the building contractor to carry out the construction project for a certain sum of money and the acceptance of that offer by

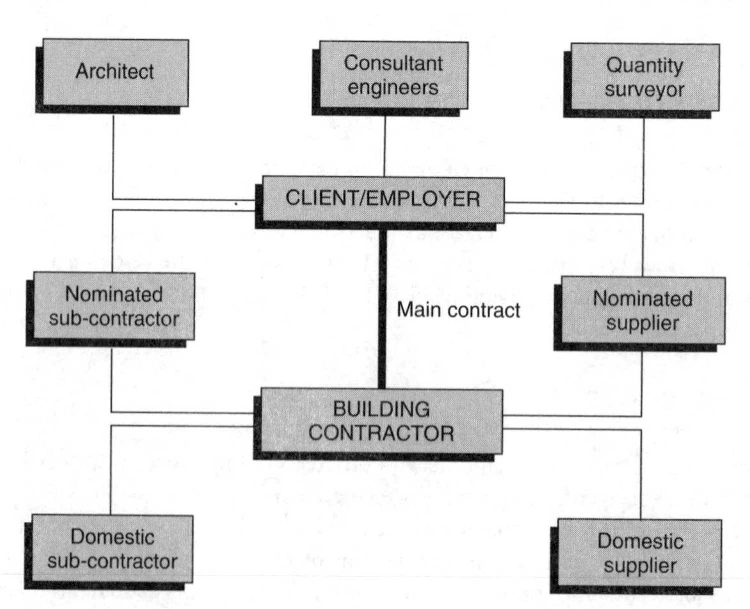

Fig. 6.1 Contractual relationship

the client. For this reason, an invitation to tender is not an offer and therefore does not bind the client to accept the lowest or any tender. The tender submitted by the contractor is the offer, but this has no binding effect if it is described as an estimate.

When a client makes an unconditional acceptance of a building contactor's offer, both parties enter into a formal contract. For small construction works such as minor repairs, the contract may be verbal or contained in a letter passing between the client and the building contractor. However, for large and complex construction projects, a formal contract is normally used. Generally, this would take the form of an agreement signed by both parties and a Schedule of Conditions setting out the obligations of the parties and matters concerning the administration of the contract and the payment of the building contractor. As the drafting of such a document requires care, legal knowledge and skill, contracting parties usually adopt one of the current Standard Forms of Contract (see section 6.10) for their contractual relationship.

6.9 PERFORMANCE AND BREACH

Usually every contractual obligation gives rise to a corresponding contractual right. Thus, where one party performs exactly and completely what it has undertaken to do, its obligations are ended and, likewise, the corresponding right of the other party is extinguished. By the same token, when all obligations which arose under a contract are discharged and all rights are thus extinguished, the contract is said to be discharged. Once a contract is discharged, the parties can no longer rely on its terms; however, they can if they so wish enforce whatever rights may arise from the discharge.

6.9.1 Discharge of contract

A contract may be discharged in any of the following ways:

1. *Performance:* A contractual obligation is discharged by a complete performance of the undertaking. However, under construction contracts, a building contractor may receive payment where a contract has been substantially performed subject to an allowance being made for defective and/or uncompleted works.
2. *Agreement:* A contractual obligation may be discharged by a subsequent bidding contract. This approach to the discharge of contract is possible where:

   (a) Neither party has completed its undertaking.
   (b) One party only has completed its undertaking and something remains to be done by the other party.
   (c) One party to contract is released by a third party who undertakes its obligations and rights (if there are any).

3. *Frustration:* A frustrated contract is a contract which has been determined prematurely due to some intervening event or change of circumstance of a fundamental nature making the contract impossible of performance. Therefore, when without the default of either party, circumstances change so that performance of a contractual obligation become radically different from that undertaken, the contract can then be termed as frustrated and thereby involuntarily discharged.
4. *Acceptance of breach:* Breach of contract occurs where, without justification, a party either fails to perform its contractual obligation expressly or by implication. When this happens, the party not in breach always has an action for damages against the party in breach and, in certain circumstances, may treat the contract as repudiated by the party in breach and refuse further performance. This therefore means that breach by one party to a contract may enable the innocent party to become discharged from further liability under it.

6.9.2 Remedies for breach

As mentioned above, if the innocent party to a repudiated contract treats the contract as discharged (provided the party has the right to do so), it is relieved from further liability. What is more, depending on the subject matter of the contract and the nature of the breach, the innocent party may sue for damages, reasonable renumeration (i.e. *quantum meruit*) or for a decree of specific performance.

Limitation periods

Under the provisions of the Limitations Act 1939, the procedural right to bring an action for breach of contract may be extinguished by the passage of time. In principle, when a party is in breach of contract, the innocent party can sue and at this moment the appropriate limitation period will begin to run. At the end of this period (six years for action founded on a simple contract or in tort and twelve years for action founded on a contract under seal), the action will be statute barred (i.e. no longer be maintainable).

A claim in respect of personal injuries for breach of duty must normally be brought within three years. In all cases, however, and as mentioned above, time runs from the date when the cause of action arose (i.e. from the time the breach of contract occurred).

6.10 STANDARD FORMS OF CONTRACT

Standard forms of construction contract are contract documents that have been produced (with a few exceptions) by the agreement of recognised bodies representing the construction industry's clients, contractors and professionals. Their purpose is to provide interested parties to a contract with suitable ready-made terms and conditions on which to contract. Various standard forms of contract are available in the UK construction industry to suit projects of almost any size and complexity and, hence, befit most methods of construction procurement. There are, for example, forms issued by the government; there are also those issued by the Institution of Civil Engineers and the Joint Contracts Tribunal. Their continued existence and use is the result of their general acceptance by contracting parties and their firm desire to avoid drafting a new form of contract each time there is a need for construction activity. Nevertheless, while it would be practically impossible to devise a standard form of contract that would take account of all eventualities that might occur on a construction project, standard forms of contract are often criticised as being a compromise containing some defective aspects in one form or another. However, the advantage from their adoption is that, with the passage of time, persons using them become familiar with their overall contents as well as their particular strengths and weaknesses.

Currently, the standard forms for construction projects in the UK are many and diverse to be usefully reviewed within the limits of this book. But generally depending on the status of client (private or public), status of designer (architect or engineer), size and complexity of project and whether bills of quantities are to be used, the client's professional advisers normally recommend the use of one of the following standard forms of contract.

6.10.1 JCT standard form of building contract (1980 edition)

This standard form of contract is issued by the Joint Contracts Tribunal and is commonly known as *JCT 80*. The Joint Contracts Tribunal represents all sectors of the UK construction industry (i.e. the construction professionals, contractors, sub-contractors and clients in the private and public sectors). Hence, the JCT 80 covers a wide range of needs and provides the basis for solving satisfactorily many problems which frequently arise in building contracts or works. Moreover, it sets out procedures to be followed for the administration of a project by the appointment of an architect (to design the building) and quantity surveyor (to produce bills of quantities and measure and value variations). In addition, the JCT 80 standard form contains several provisions, including:

- Extensions to the project's completion time.
- Powers of the architect to issue instructions and to vary the works.
- Insurances required by the client and contractor.
- Level of compensation required by the client for late completion by the contractor.
- Extra payment required to cover the contractor's direct loss and/or expense where the cause is the responsibility of the employer or his/her professional advisers.
- Reimbursement for increases in the cost of resources during production (normally referred to in the contract as fluctuations).

Owing to the form's comprehensive provisions, it is used for larger and complex projects which are fully designed and detailed for single-stage selective tendering.

The advantages and disadvantages of the use of JCT 80 may be summarised as follows:

Advantages

- It provides comprehensive coverage and, hence, contains clauses or conditions that are able to address many construction problems.
- It has a range of ancillary documents/special supplement to amplify the terms and conditions it contains.
- It is widely accepted and most generally used and, hence, most of its users are familiar with its contents as well as its strengths and weaknesses.
- It recognises the contractual anomalies arising from nomination of sub-contractors and provides adequately for them.

Disadvantages

- It is lengthy and too complex to comprehend all its provisions fully.
- Its provisions for nominated sub-contractors are lengthy and complicated.
- It contains uncertain terms such as direct loss and/or expense, practical completion and *force majeure* which, on the surface, appear to be quite clear and certain until a dispute arises.
- It is cumbersome and too often opens the ground for dispute by its inability to provide unequivocal answers on the rights and obligations of the parties.

Generally, the JCT 80 is produced in various versions. These are '*with firm quantities, approximate quantities or without quantities*' and each of these has a version for use by private clients or local authorities. There are, however, a small number of differences between the private and local authority editions. At the time of writing, the JCT 80 had been published in six versions.

6.10.2 JCT IFC 84 standard form of contract

The IFC 84 is the Joint Contracts Tribunal Intermediate Form of Contract for building works of simple content. This JCT form of contract was first produced in 1984 and, as the name implies, it is less complex than JCT 80. This standard form is intended to bridge the gap between the minor works form (MW 80), which is suited to smaller projects, and JCT 80. The IFC 84 form is suitable for use on building projects which are:

- Medium in size with simple work content involving normally recognised basic trades and skills of the construction industry.
- Without any complex service installations or other specialist work.
- Adequately specified and/or billed as appropriate prior to invitation to tender.

The IFC 84 form has no provisions for nominated sub-contractors. However, like JCT 80, it provides for the appointment of an architect and a quantity surveyor for the design and administration of a construction project. It also contains adequate provisions for contractors to recover fluctuations in the cost of resources.

The advantages and disadvantages of the use of IFC 84 standard form of contract may be summarised as follows:

Advantages

- It is very much simpler than JCT 80 and benefits from precise cross-referencing.
- It is comprehensive and in most cases may be suitable for somewhat large contracts provided the work is of a reasonably simple nature.
- It is flexible and contains adequate provisions for contractors to recover increases in the cost of resources during production and, hence, it can be used for construction projects with a long production time.

Disadvantages

- It is not suitable when complex specialist works form a large part of the building project.
- It is not generally suitable for large and complex projects with a long production time.
- It cannot be used on projects in which the contractor will be wholly or partly responsible for design.
- It cannot be used on projects in which there is a necessity for nominated sub-contractors and nominated suppliers.
- It has complex sub-contract *naming* terms for sub-contract work (whereby the architect selects the sub-contractor but, once appointed, the sub-contractor is treated as the contractor's domestic sub-contractor).

6.10.3 JCT MW 80 standard form of contract

JCT MW 80 is the Joint Contracts Tribunal Agreement for Minor Building Works (currently 1980 edition). This form of contract is suitable for smaller projects (minor building or maintenance work) based on specifications and drawings for an agreed firm lump sum. In addition, it provides for the appointment of an architect and a quantity surveyor, but does not recommend the use of bills of quantities and nominated sub-contractors.

The advantages and disadvantages of the use of MW 80 standard form of contract may be summarised as follows:

Advantages

- It is short, simple and easy to follow.
- It is flexible and condenses all relevant points of the main form of contract (i.e. JCT 80).

Disadvantages

- It has inadequate provisions for direct loss and/or expense claims.
- The form is not appropriate for works for which bills of quantities have been prepared or in which valuation of variations requires the use of a schedule of rates.
- The form contains inadequate provisions for extensions of time and insurance.
- It is unsuitable for prime cost work and hence, when in use, nominated or named sub-contractors cannot be employed.

6.10.4 ICE standard form of contract

The full title for this standard form of contract is the Conditions of Contract and Forms of Tender, Agreement and Bond for use in connection with works of Civil Engineering Construction. It is currently in its 6th edition. This standard form has been prepared and issued jointly by the Institution of Civil Engineers, the Federation of Civil Engineering Contractors and the Association of Consulting Engineers.

The ICE Form, as this standard form of contract is generally known, is used for civil engineering projects and is suitable for projects for both private and public sector clients. Dependent on the size of project, the form provides for the appointment of an engineer, a resident engineer or assistant engineer or clerk of works to supervise and inspect works in progress. Owing to the uncertain nature of civil engineering works, all bills of quantities prepared for civil engineering projects are known to contain provisional quantities and hence, the actual work is remeasured on completion. For this reason, the form does not provide for the statement of a specific cash sum for which the successful contractor will undertake the works; rather it stipulates the use of the terms *Tender Total* and *Contract Price* as the sum to be ascertained in accordance with the provisions of the contract. Furthermore, the form does not provide for nomination of sub-contractors and, hence, a client does not have to worry about them in his or her projects.

The advantages and disadvantages of the use of use of ICE standard form of contract may be summarised as follows:

Advantages

- The ICE standard form is comprehensive and, moreover, the wording of the conditions are not ambiguous.
- It recognises the high risk factor in civil engineering projects and the economic sense of allocating risks between employer and contractor. For this reason, when in use, the contractor does not bear all the contractual risks and responsibilities.
- The engineer's periodic intervention during production ensures safe execution of the works as well as the safety of all those engaged upon it.
- The contractor can claim successfully for a troublesome physical site condition which he or she could not have reasonably foreseen at the time of tender.

Disadvantages

- Measure and value rather than lump sum contract means that the client may not know the total financial commitment at the outset.
- The conditions are drafted to suit large-scale civil engineering work with the emphasis on ground works and the relatively high cost of plant and temporary works and is, hence, unsuitable for building works associated with civil engineering works.
- The engineer's control over production method, sequence outlining and authority to decide what is required and to issue instant variations may interfere with the contractor's production programme.

- The engineer has considerable powers and, as it is his or her responsibility to explain and adjust any ambiguities or discrepancies in the contract document, the contractor stands to lose if the engineer displays partiality in the interpretation or adjustment.

6.10.5 The New Engineering Contract

The New Engineering Contract (NEC) was first published as a consultative document by the Institution of Civil Engineers in 1991 and was published as a standard form of contract in 1993. The complete contract contains core clauses, schedule of options, option clauses (main and secondary), schedule of actual cost and schedule of contract data. The intention of the authors of this form was, among other things, to provide a flexible standard form of contract capable of being used for both building and civil engineering projects. It is a new and innovative standard form of contract which facilitates teamwork by giving the employer choice of contract strategy and allocation of risk. It caters for the proliferation of disciplines and is also versatile to span the greater spectrum of contract activity without risking contractual disputes. Hence, the authors contend that use of the NEC standard form of contract will certainly lead to the achievement of the client's project objectives in terms of its ultimate quality, performance, cost and completion time. Furthermore, its adoption is likely to benefit contractors in the form of increased profits.

The advantages and disadvantages of the use of NEC standard form of contract may be summarised as follows:

Advantages

- The pre-estimation of the time and financial effects of instructions enables the client to know the overall position of his or her project as a result of an engineer's instruction.
- Subject to the acceptance of and agreement to a contractor's quotation, the client can request the acceleration of the project at any time during production on site.
- The client will be able to recover damages from the contractor if an installed equipment fails to function in accordance with a specified performance level.
- The NEC standard form of contract provides for detailed procedures for managing construction project risks as they occur.
- The increased powers of an adjudicator limits the use of arbitrators in the solution of contractual disputes and this encourages parties to settle their contractual disputes amicably.

Disadvantages

- Although the intention of the authors of the NEC standard form of contract is to make the contract simple, the opposite is achieved as many of its clauses are obscure.
- Owing to the obscurity of the meaning of some clauses, their diverse legal interpretation is likely to open the floodgates to contractual claims.

- The NEC standard form of contract is silent on some important points such as revised dates for completion and the action required should a project manager decline to consent to the issue of instructions to change illegal or impossible work.

6.10.6 GC/Works/1 standard form of contract

The GC/Works/1 is the General Conditions of Government Contract of the Building and Civil Engineering Works. It is issued by the Department of the Environment (DoE) and the current issue is the 1989 edition. This document was unilaterally prepared by the contracts coordinating committee on behalf of the DoE and is therefore in use by the central government departments for both building and civil engineering projects. Generally, however, the form is not suitable for use by other public bodies and construction clients in the private sector.

The GC/Works/1 standard form of contract provides for the appointment of a superintending officer who can be an architect, a civil engineer or a quantity surveyor to supervise the client's construction project. In addition, the form provides for the appointment of a resident engineer or a clerk of works who is vested with the authority to inspect works in progress and test material as the superintending officer may require.

The advantages and disadvantages of the use of GC/Works/1 standard form may be summarised as follows:

Advantages

- The document is good in terms of greater simplicity and precision and, hence, minimises unmerited contractors' claims.
- The contract gives the client, in all other respects, an incomparable better protection than other standard forms of contract.

Disadvantages

- The form is unilaterally drafted and, hence, is more in favour of the client. Moreover, from the contractor's point of view, it is considered to be an onerous form of contract.
- The supervising officer has considerable powers and as his or her decisions are final and conclusive (and not subject to arbitration), this is considered to be unfair to the contractor.
- The client has special powers to determine the contract at any time (not necessarily following any default by the contractor) but there is no provision allowing the contractor to act in a similar manner.
- The client has power to recover any losses suffered from any other contract placed with the contractor.

6.10.7 PSA/1 (with quantities) standard form of contract

PSA/1 is the Property Services Agency's General Conditions of Contract for Building and Civil Engineering works (currently 1994 edition). It is a standard form of contract which was first published by HMSO (Her Majesty's Stationery Office) in August 1994 for use on construction projects. This standard form of contract contains some provisions which are absent from other published standard forms of building and civil engineering contracts. Some of these are:

- Pre-agreement of prices of variations.
- Agreement of construction programme prior to acceptance of tender.
- Stage payment method of assessing total value for works in progress.
- Specification of the desired level of design liability and contactual risk.
- Incorporation of project manager's role as the central contractual personality.

This standard form also provides the framework for a quicker settlement of final account, contractual disputes and for regularisation of interest and finance charges.

The advantages and disadvantages of the use of the PSA/1 standard form of contract may be summarised as follows:

Advantages

- The document is comprehensive and caters for most problems that may arise in construction contracts.
- It provides a faster method of settling the final account, contractual claims and contractual problems generally.
- It provides the client with a single point of contact.
- The level of design liability is clearly defined
- It provides the contractor with an incentive to help develop the construction project.

Disadvantages

- The document is bulky for anyone to be fully versed with and apply its provisions within a short period of time.
- The stage payment rule is likely to deplete the contractor's cash on stretched construction activity.
- The document is fairly new and untried, and hence its weaknesses have yet to surface.
- Contractors have problems of their own and should be left alone to press on with the construction organisation and production processes (instead of involving them in measures required for the achievement of the client's project cost objectives).

6.11 SUPPLEMENTS TO STANDARD FORMS OF CONTRACT

As pointed out above, it is essential that the use of any of the above standard forms of contract is specified or made known to the bidding contractors at tendering stage so that an appropriate financial allowance may be made in their bids to cover any risks associated with their adoption. Furthermore, where required, there will be a need to include within the foregoing standard forms of contract in respect of the following:

6.11.1 Bonds

A bond is simply a contract of guarantee. It takes the form of a written undertaking by a guarantor (i.e. the surety company – a bank or insurance company) to accept responsibility for the performance of a contractual obligation. A particular type of bond found in the construction industry is known as a performance bond, which is a protection against failure to complete the construction project.

As mentioned above, bonds are normally contracts of guarantee or indemnity provided by banks or insurance companies for the due performance of a contractor's obligations under his or her contract with the client. A bond usually constitutes a promise by the surety company to pay a sum of money – commonly limited to 10 per cent of the original contract sum – to the client if the contractor fails to meet his or her contractual obligations. The consequence is that if the contractor defaults in the performance of his or her contractual obligations, then the surety company makes payment up to the level of the stated sum.

Generally, as a rule, a performance bond remains in force until the stated discharge date and, in construction projects, this date is usually either after practical completion of the works or after making good any defects. However, should the practical completion or making good of defects occur earlier than the bond date, the bond cannot be recalled or withdrawn unless the client agrees to an earlier release date. Furthermore, a performance bond is not an insurance policy which normally is a contract of indemnity under which the insured is indemnified in the event of loss, subject to the adequacy of the sum insured. Moreover, there are three parties under a performance bond (i.e. the contractor, the client and the surety company) as opposed to two under an insurance policy (i.e. the insurer and the insured). Once a bond is issued, it cannot be cancelled until the stated discharge date or until the subject matter of the indemnity has been completed satisfactorily, however, an insurance policy can be cancelled before its expiry date.

Types of bond available

There are currently two types of bond in use in the construction industry. These are the conditional bond and on-demand (unconditional) bond.

1. *Conditional bond:* Under conditional bond, a fault, responsibility and loss will be required to be proved or established before the requisite money is paid out. For this reason, it is sometimes referred to as default bond. Under this bond, the surety company will usually require some detailed collaboration and substantiation of the client claim and, if not satisfied, it can withhold payment until compelled by a Court judgment to pay.
2. *On-demand (unconditional) bond:* On-demand bonds are unconditional instruments, issued by banks, that contractors have to give to their clients. These type of bonds are generally payable without proof of default or any

damage/loss sustained by the client. In simple terms, on-demand bonds call for payment of the sum '*upon first written demand*' to the bank without any proof of default. The client who obtains such a bond has some protection against a defaulting contractor. It may only be 10 per cent of the contract sum, but the attractive element for the client is that no questions are asked when the bond is demanded. Nevertheless, the client is required to make the demand in the appropriate form, served at the right time and at the right place, especially where compliance with such express procedural requirements have been specified in the bond.

Practical considerations

1. *Clarity of bond wording:* The cost of a bond is normally added to the contract sum. Hence, when the client's professional advisers recommend the need to have a bond from the contractor, the client must specify the precise wording of the bond that is required to be given by the contractor. This action is crucial as lack of clarity of what is required can create problems in collecting the bond in case of contractor's default. Moreover, a reference should be made to the need for this guarantee in the contract documentation.
2. *Parent company guarantees:* Where the contractor is part of a larger group of companies, the client's interest may be secured by asking the parent company to provide a guarantee against defects or failure to perform by the subsidiary company. This guarantee is only as secure as the parent company that gives it, and if the whole group of companies collapse, then the parent company guarantee will provide no security at all. For this reason, it is better for the client's professional advisers to insist on a guarantee from the ultimate holding company of the contractor and not from an intermediate company. On grant of the guarantee, the parent company stands in for the contractor and, if the contractor defaults, the parent company must meet any liabilities that would otherwise have attached to the contractor. However, unlike a bond, a parent company guarantee costs the contractor no money and, hence, the client incurs no additional cost.

6.11.2 Collateral warranties

Collateral warranty is a new development whereby, as a result of the client or the ultimate users demand for contractual commitment on which to sue when the need arises, the construction team are required to give a warranty to them and their successors. The essential purpose of a collateral warranty is to extend contractual rights to a third party who is not a signatory to a particular contract. The client usually ask for these warranties prior to making appointment of his or her professional advisers. But in the case of a construction contract, this can be done by incorporating various forms of warranty into the contract documentation with appropriate provisions setting out clearly when and from whom warranties can be requested and to whom they must be given in those circumstances. There are no fixed rules generally but, for example, specialists who are responsible for design, such as mechanical or electrical contractors, may be required to give a warranty for the quality of their design for up to (say) fifteen years. During this period, they can be sued by the client or a new owner (where the building is sold) who claims that equipment or a component designed is not fit for its intended purpose. For this reason, the warranty is a time bomb and its adequacy places the client in a good sale, letting and owner-occupier position.

6.11.3 Retention

Retention is the sum of money deducted or withheld by the client from the amount of money due to the contractor for work satisfactorily completed. The percentage used to calculate the amount of deduction is called *the retention percentage* and this percentage is stated in the contract conditions and, hence, known to contractor at tender stage. For this reason, the full retention percentage is deducted from the contractor's total interim valuation for work that has not reached practical completion. However, one-half of the retention is released when the works reach practical completion and the remainder released after the issue of an architect's certificate of completion of making good defects.

The purpose of deducting or withholding part of the total value of work satisfactorily completed include:

- Ensuring that the works will be properly completed and that all defects in the work will be made good.
- Provision of an incentive for the contractor to proceed diligently and complete the works promptly.
- Provision of protection to the client against the effect of the contractor defaulting.
- Serving as a source where the client can deduct monies for:

  – Recovery of liquidated and ascertained damages.
  – Cost of employing others to carry out variations or to correct defective work.

- Acting as a cushion and defraying some of the additional costs when the contractor fails to complete in the case of determination or liquidation.
- Enabling a client to pay the nominated sub-contractor and suppliers direct should the contractor fail to pay them.

The retention percentage is normally between 3 to 5 per cent. However, to provide the client with additional security, this figure can be as high as 10 per cent. But the effect of a higher retention percentage affects the contractor's cash flow and, moreover, increases tender figures as its cost form a part of all tender figures. Generally, the retention is trust money for contractors, and to safeguard their interest contractors have the right to ask the client to place the retention money in a separate bank account. The retention money so placed protects contractors from financial loss should the client go into liquidation.

6.11.4 Liquidated Damages

In order that the client may secure his or her development project by a definite date, a clause is usually inserted in a contract to make the contractor liable for payment of an amount should he or she overrun the completion date on the contract. As the amount so inserted in the contract is known at the time of entering into contract, it aims at determining in advance the extent of future liability for a specified breach and motivating the contractor to work diligently to meet the project's programme. It must also be remembered that liquidated damages are not a penalty (which is payment to make an offending party suffer for his or her faults) but rather it is a genuine pre-estimate of damage which may be suffered from breach of contract. Therefore, the amount of liquidated damages should not be excessive by comparison with the likely greatest loss that the client would suffer from breach of contract. If the sum stated as liquidated damages in the contract is too high, it may be construed by the Courts as a penalty and hence be unenforceable. On the other hand, the amount of liquidated damages may be agreed in advance with a contractor as representing the loss the client is likely to suffer in case of breach in order to avoid the possibility of the amount being construed as a penalty. When the contractor becomes liable to pay the amount of liquidated damages provided in the contract, the sum is deducted from monies due to him or her as they arise under interim valuations. Nevertheless, any such deduction is at the discretion of the client, who decides when and how to recover the money. If the client so wishes, he or she may waive any right to damages, in which case no deduction is made. However, when liquidated damages become payable, it will be prudent on the part of the quantity surveyor to remind the architect of the need to deduct monies due to the contractor from interim valuations. On receipt of this information, the architect will be in a position to advise the client accordingly when the certificate is issued.

6.12 TRADITIONAL CONTRACTUAL ARRANGEMENT

At a certain stage in the client's construction project his or her professional advisers adopt a suitable contractual arrangement for the project. Generally, contractual arrangements set out the legal relationship parties wish to establish and, hence, creates rights, obligations and procedures for resolving contractual disputes. In addition, contractual arrangements in the construction process establish the basis for making payment to the contractor. The factors which influence the choice of appropriate contractual arrangement include:

- Size, nature and complexity of development.
- Dates for commencement and completion.
- The ability to define the client's requirements clearly before contract.
- Adequacy of construction information on which to establish client's cost limit.
- Availability of valid and adequate construction information on which to obtain tenders.

- The scale of changes the client is likely to effect during the construction phase.
- State of the national and international economies and their effect on the construction market.

6.12.1 Types of contractual arrangement

The choice of an appropriate contractual arrangement is of fundamental importance to the successful completion of a construction project. For this reason, the type of contractual arrangement selected should be incorporated in the contract documentation so that tenderers are aware of the contractual arrangement proposed and their contractual obligations. The various types of contractual arrangements available to construction projects are as follows:

Lump Sum Contract

Under this contractual arrangement, the contractor consents to execute the entire work described or specified for a stated total sum. The agreed sum is normally based on information derived from drawings, specification, bills of quantities and/or site inspection. To arrive at the pre-estimated price, the contractor takes into account all contractual risks involved, the condition of the construction market and his or her current workload. The pre-estimated price is paid to the contractor regardless of the actual costs incurred in executing the works, providing there are no variations.

As a rule, lump sum contracts may be procured on either fixed or fluctuating price. Financial adjustments to the contract sum, reflecting changes in labour and plant rates or material prices during the progress of the works, is permissible in the former but not permissible in the latter, apart from statutory fluctuations. Furthermore, lump sum contacts may be procured either on drawing and specification (for minor works, maintenance and specialist works) or drawings and bills of quantities (for all work except minor works). It may be adopted also for projects where there is adequate information to enable the client to know his or her total financial commitment before contract.

Continual contracts

When the client wishes to pursue a programme of construction work, his or her professional advisers may recommend one of the following contractual arrangements to effect a saving in tendering cost and expedite production.

1. *Serial contracts:* Under this contractual arrangement, the contractor undertakes to enter into a series of separate lump sum contracts in accordance with the terms and conditions set out in an initial offer. The standing offer may be determined by competitive pricing of a notional bills of quantities containing key items of the proposed construction projects.
2. *Continuity contracts:* Where the client wishes to obtain the benefits which arise from continuity of work, a building contractor may be asked to enter into

negotiations based on an original lump sum contract. Once the negotiations are completed and an agreement reached, the contractor executes the construction works as separate contract packages within their own parameters.

3. *Term contracts:* Under this contractual arrangement, the client commissions a building contractor to undertake specified construction work within a defined cost limit for a definite period, often of twelve to thirty-six months' duration. The valuation of the contractor's work is priced on either a schedule of rates or a cost reimbursement basis. This type of contractual arrangement is suitable for low-value on-going repair and maintenance work where contractors submit invoices for payment on completion of the specified works.

Measurement contract

Price for sections of construction work under this contractual arrangement is pre-estimated but the total price cannot be ascertained until the work is measured and valued on completion. The evaluation of the measured construction work is by the application of an agreed unit rate obtained from either bills of quantities or schedule of rates. This contractual arrangement can be procured on an approximate bills of quantities (when client's requirements are not known in advance) or schedule of rates (when the client's requirements are insufficient to permit the production of bills of approximate quantities). It may be adopted for projects where the client's requirements are not clearly defined or where prompt commencement on site is required.

Cost reimbursement contracts

Under this contractual arrangement, the client undertakes to pay the contractor the prime cost (i.e. the actual cost of labour, plant and materials utilised in the execution of the construction works). In addition to the prime cost, the contractor is paid an agreed sum to cover profit and establishment charges. This contractual arrangement may be adopted for projects where:

- The client may wish to influence the execution of the works and, hence, assume the entire risk of site operations.
- An early start is required but the extent of the works cannot be accurately predicted.
- A high standard of work is required.
- Work is of an emergency, repair and experimental nature.

This contractual arrangement may be criticised for lack of financial incentive which would otherwise encourage contractors to perform efficiently. However, the following variations may be introduced to motivate and enhance the contractor's site performance.

1. *Fixed fee:* Under this arrangement, the contractor is paid an agreed fixed fee based on the estimated cost of the construction work. This induces the

contractor to work efficiently for profit. However, adequate pre-contract information is required for the preparation of a project's estimate on which to agree a fee, but this may be lacking.

2. *Target cost:* Under this method, an estimate is produced for the construction project and, once this estimate is agreed, it becomes the target price which establishes the basis for the determination of a fee for overhead charges and profit. On commencement of site production, the contractor is paid the prime cost and the fee, which is adjusted to correspond with the increase or decrease of prime cost over the target cost. The main problem with this approach lies in the agreement of a realistic target and the effects of costly variations which eventually take the construction cost over the agreed target price.

6.13 FIRM OR FLUCTUATING PRICE CONTRACT

Most construction projects are procured on either firm price or fluctuating price contract. For this reason, at the pre-contract stage, the client's professional advisers decide on a form of contract appropriate for the client's construction project. Firm price contracts are those where a contractor's claim for reimbursement of changes in costs is limited to those relating to statutory contributions, levies and taxes. Usually, when a project is placed on a firm price contract, the contractor is asked to allow for all increases or decreases in costs of labour, material and plant that occur during the production phase. (Further discussion of this has been provided in the author's *Understanding Tendering and Estimating*.) Fluctuating price contracts (i.e. the alternative to firm price contracts) are those where the contractor can claim for reimbursement of changes in cost to cover a wide range of labour, material and plant costs as well as statutory contributions.

The design team's decision on the selection of either firm price or fluctuating price contract is normally influenced by the length of a contract period and the current and forecast rate of inflation. If a construction project is of short duration (i.e. not exceeding two years with stable inflationary pressure), then a firm price contract may be proposed. However, where the contract period is long with uncertain inflationary pressures, then a fluctuating price contract arrangement becomes more appropriate. If this approach were not adopted in an inflationary economic climate, all bidding contractors would, spontaneously, inflate their bid prices to uneconomic levels to cover this uncertainty.

SUMMARY

Owing to the risks and large sums of money involved in construction projects, all agreements need careful drafting to enable parties to understand their respective rights and obligations under the contract. The law regarding contracts is intricate in nature and, hence, legal agreements are not easy to draft. However, in the construction industry, there exist several ready-made standard forms of contract suitable for use on all kinds of projects. This saves parties to various construction projects time and money in drafting new agreements to suit their particular contract. Moreover, most of these ready-made standard forms of contract have been in circulation for several years and most construction clients and their professional advisers are familiar with their usage as well as with their weaknesses and strengths.

FURTHER READING

Ashworth, A. 1988 *Cost studies of buildings.* Longman Scientific and Technical.
Elliot, R.F. 1988 *Building contract litigation.* Longman.
Fumston, M.P. 1986 *Law of contract.* Butterworth.
James, M.F. 1994 *Construction law.* Macmillan.
Keating, D. 1995 *Keating on contracts.* Sweet & Maxwell.
Major, W.T. 1977 *The law of contract.* M&E Handbooks.
Murdoch, J.; Hughes, W. 1993 *Construction contracts – law and management.* E&FN Spon.
Seel, C. 1984 *Contractual procedures for building students.* Holt Rinehart Winston.
Turner, D.F. 1977 *Building contracts: a practical guide.* George Godwin.
Uff, J. 1991 *Construction law.* Sweet & Maxwell.

CHAPTER 7

CONTRACT DOCUMENTATION

7.1 INTRODUCTION

A document may be described as anything on which marks have been made with the intention of communicating information to a third party. Hence, such things (on paper) as writing, drawing, printing, typescript, photograph and the like are documents. The documents which are brought together to form the evidence of a contract, legally agreed by the parties and therefore signed as such, are termed *contract documents*. Contract documentation, on the other hand, is the recording of the content, terms and conditions in the formal documents of the contract between the parties. In construction contracts, this embodies the entire written and drawn construction information. Contract conditions also constitute the core of the agreement between client and contractor in the construction process. They govern the erection of a proposed building with an undertaking by the contractor for a consideration (by way of renumeration), to construct the project for the benefit of the client. Hence, if at any time during the progress of the works, a contract condition is broken by one party, the other party may repudiate the contract and elect to treat the contract as at an end and sue for damages.

Contract documentation plays several distinct but important functions in the construction process and these may be summarised as follows:

- It delineates the works in terms of drawings (which show position and extent) specification and bills of quantities (which describe quantity, quality and position).
- It shows clients the extent of the construction project and gives an indication of the financial and legal obligations before contract.
- It reflects the intention of the parties and places rights and obligations on them.
- It creates a situation which enables building contractors to bid for jobs on the same information and terms without any ambiguity.
- It is a record of the scope of the work in terms of price, quality, time, risks and determination of disputes.
- It provides a fair, equitable legal framework which ensures that work is carried out in a proper manner and that building contractors receive payment for work satisfactorily completed.

- It can be submitted as evidence to establish a point in dispute (i.e. failure to comply with rights and duties set out in the contract documents).
- It can be produced in order to prove the existence of a collateral agreement or warranty between parties or among project participants.

The exact work involved in the preparation of a contract documentation varies in accordance with the contractual arrangement selected to suit a client's project strategy. However, to perform the above functions properly and to safeguard the client's interest, a contract documentation should explain the full bargain between client and building contractor. In addition, it should adequately describe the scope, quantity, quality and position of the work and define the rights and obligations of parties. A carefully drafted contract documentation promotes smooth running and successful completion of a construction project. Also, to prevent disputes arising during the production phase, contract documentation should be carefully, adequately and accurately prepared and be consistent throughout. This is an important point as post-contract project information is not intended to impose any additional obligations beyond those contained within the contract conditions. The purpose of post-contract project information is to clarify or supplement the existing information and, therefore, an adequately prepared contract documentation achieves the above objective satisfactorily.

7.2 COMPOSITION OF CONTRACT DOCUMENTATION

The nature of contract documentation is diverse, serving different purposes within the construction process, but a typical construction contract documentation may comprise the following:

7.2.1 Agreement or articles of agreement

The articles of agreement constitute the actual contract between the parties, and in a construction project it is the contract between the client and the contractor. This usually sets out the date and the parties; defines the works; defines the responsibility for the preparation of drawings and bills of quantities; defines the contract sum, payment times and methods; and defines the resolution of disputes. Although the architect and the quantity surveyor are not parties to the contract, by virtue of the important services they provide they may be named in the articles of agreement.

7.2.2 Conditions of contract

The conditions of contract set out the obligations and rights of the parties and detail the conditions under which the contract is to be carried out. Generally, they attempt to provide for the various problems which can arise during and after execution of the construction project. In the UK construction contracts generally, the identification of conditions of contract is frequently contained within various standard forms of contract such as the JCT (Joint Contracts Tribunal 1980), the IFC (Intermediate Form of Contract 1984) and so on. However, the quantity surveyor sometimes drafts additional contract clauses in the preliminaries section of the bills of quantities in an attempt to govern certain contractual matters not covered by the standard contract conditions. At times, too, the quantity surveyor may make amendments or deletions to contract conditions in the standard form of contract to suit a particular project. However, in making such amendments, the quantity surveyor should make certain that ambiguities, discrepancies or contradictions are avoided and ensure that it can withstand close scrutiny in a Court of Law or in arbitration.

7.2.3 Drawings

Drawings provide fundamental information for construction projects and, hence, are used as a means of communication. Generally, drawings provide information defining shape, size and location of a construction project. In addition, drawings, in conjunction with schedules, indicate the materials to be used on a construction project. The number of drawings produced depends on the size and complexity of project; however, initially the architect commences by producing the general arrangement drawings which are normally used by other consultants and specialists for setting out dimensions, preparation of cost plans and for seeking planning approval. After this stage, the architect proceeds to the production of large-scale details and junctions. As a medium for communication, drawings must be clear and be able to inform the users of all constituent parts of the construction projects, composition of materials to be used and jointing and fixing methods. Every drawing should also show its title, sheet number and the scale to which it is drawn to facilitate retrieval, cross-referencing and general use.

7.2.4 Specification

Specification is a contract documentation which either describes the designer's particular requirements or a specific level of performance for a construction project. Specifications may therefore be discussed under two headings, as follows:

Material/workmanship specification

Material/workmanship specification is a contract documentation produced to supplement drawings in conveying construction project information. Generally, except in very small projects, drawings cannot incorporate all the information required by the users to assess the cost, particular workmanship and details involved. Hence, a specification is produced to provide such additional information as special construction problems, standard of workmanship expected, preparation to surfaces and components, standards of materials to be used, design mixes of concrete, relevant British Standard (BS) codes of practices and so on. In effect, a specification is a contract documentation which defines in words alone the nature and extent of the works shown on contract drawings.

Specifications may also be produced as a contract documentation in their own right. This is particularly so in small construction projects which does not adopt the use of bills of quantities and, hence, the specification and drawings form the contract documentation. In such situations, a schedule of descriptions with or without quantities is prepared to provide the basis of price determination, valuation of variations to design and preparation of interim valuations. On the other hand, specifications can stand alongside or even be absorbed by bills of quantities.

Performance specification

A performance specification is a contract documentation in its own right and deals with a specific level of performance (i.e. it calls for results and prescribed service behaviour on completion of a component or system without stipulation of methods) rather than the achievement of a level of workmanship. In effect, a performance specification is concerned with what is needed rather than how the need is attained. As a contract documentation, a performance specification may be used on construction projects of any size, any form of construction and, can even stand alongside or be combined with bills of quantities. A typical performance specification is composed of the following sections:

1. *Preliminaries:* This is virtually identical to the preliminaries section of a typical bill of quantities.
2. *Contract conditions:* This is identical to that of contractor designed contracts.
3. *Specification:* This is the actual performance specification which specifies the performance required for individual parts, sections or elements of the structure, but without quantities.

In the execution of a construction project under a performance specification, the contractor contracts to furnish a system or component possessing the specified characteristics and he or she warrants its future performance in service. The contractor is thus given a wider choice of materials and at times details of assembly with respect to the part or element under consideration.

7.2.5 Bills of quantities

The bills of quantities is a form of contract documentation which contains a schedule of fully described and quantified items of labour, plant, materials and other works set down in a systematic, recognised manner. In order to achieve a uniform method of measurement format and layout, the bills of quantities are prepared in accordance with an agreed standard method of measurement. The preliminaries section of the bills of quantities contains particulars of the type of contract in use and details of general matters which affect the whole project. The measured section contains items of work described and quantified in a manner to facilitate pricing. These are grouped in appropriate sections (e.g. in trades, elements, operations). The end of the bills of quantities contains the general summary to which the section totals are transferred for computing, and this figure is the contractor's bid for executing the project. (Example 7.1 shows a page from the bills of quantities for a typical construction project.)

Example 7.1 Sample page of bills of quantities

					£	p
	F MASONRY					
	F10 Brick/Block Walling					
	Facing brickwork Type A: in sulphate resisting mortar mix D1					
	Walls					
A	Half brick thick.	38	m^2			
B	One brick thick.	147	m^2			
	Facing brickwork Type A; in coloured mortar mix A3; finished with a bucket handle joint; as clause F121					
	Walls					
C	Half brick thick; facework one side.	46	m^2			
D	Extra; plinth header (type 2.5.2.2); facework one side.	5	m^2			
	Blockwork; as clause F127; in ordinary mortar mix D1					
	Walls					
E	100 mm thick.	53	m^2			
F	215 mm thick.	73	m^2			
	F30 Accessories/Sundry Items for Brick/Block/ Stone Walling					
	Damp-proof courses; as clause F306; bedding in mortar mix D1					
G	Not exceeding 225 mm wide; horizontal.	47	m^2			
H	Over 225 mm wide; vertical.	51	m^2			
J	Over 225 mm wide; horizontal.	85	m^2			
	Cavity Trays					
K	Over 225 mm wide; horizontal.	12	m^2			
			To Collection	£		

Besides being part of the contract documentation and assisting in competitive tendering and post-contract cost management, bills of quantities lend themselves to many other uses, including:

- Provision of uniform construction project information on which tenderers base their bids.
- Facilitation of a periodic preparation of valuation for interim payments.
- Provision of the basis for the valuation of variations that occur during the progress of the works.
- Provision of useful information for use by:
  - Quantity surveyors in estimating, cost planning and cost reporting.
  - Quantity surveyors in the preparation and evaluation of contractual claims.
  - Contractors for job planning and materials/components scheduling.
  - Architects for administering the site operations and progress/quality inspections.
  - Courts of law as evidence in arbitration or during settlement of contractual disputes.
- Facilitation of procurement of sub-contractors' quotations for various sections of work.
- Constitution of the basis for negotiating further contract (e.g. separate or continuation contracts).
- Provision of guidance to the contractor when ordering materials. Also, as it gives an itemised list of component parts of the building, the contractor can check the quantities with that contained in the contract bill and reconcile any differences.

7.3 PRODUCTION OF BILLS OF QUANTITIES

The production of bills of quantities is a process in which a quantity surveyor translates drawings into separate areas of work or sequences of work. This operation is laborious and time consuming. Many items are measured in metres and may be cubic (m^3), square (m^2) or linear (m). Some items are also enumerated (No *or* Nr) while others such as structural steelwork and steel reinforcing bars are measured by the tonne. A sample page of a bill of quantities produced under the standard method of measurement (SMM7) common arrangement of work section is given in Example 7.1 above. The quantities for each item are presented in a single total irrespective of location on the project, and financial totals on pages are collected in a summary to ascertain the total value of the works.

7.3.1 Traditional method of bill production

The traditional method of preparing bills of quantities is in the following sequence of operations:

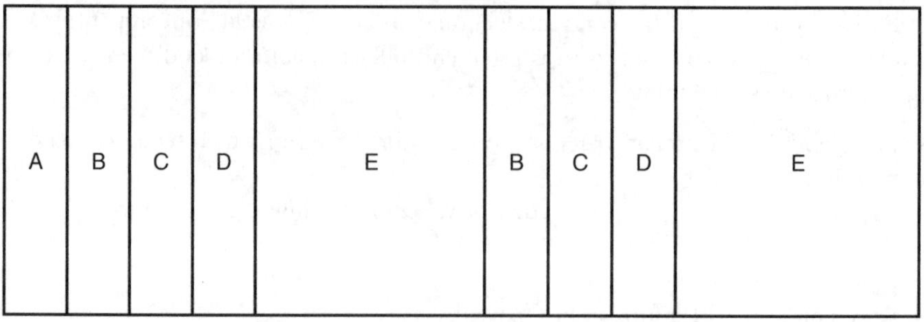

A	B	C	D	E	B	C	D	E

A = Binding column

B = Timesing column – where multiplying figures are set against description of an item of work that occurs several times within the measurement.

C = Dimension column – where the sets of dimensions relating to the description of items of work are entered. These may be lineal, square or cubic.

D = Squaring column – where sets of dimensions are multiplied out.

E = Description column – where the descriptions of the items of work are entered; it may also be used to set down preliminary calculations.

Fig. 7.1 Sample of dimension paper

Taking off

Taking off is the measurement of all sections of building work in accordance with an agreed standard method of measurement of building works. The person who carries out this operation is traditionally known as a *Taker Off*. The taker off normally enters the dimensions read or scaled from drawings in recognised form on specially ruled paper known as *Dimension Paper* (see Figure 7.1).

Working up

The process of working up comprises the following stages of activities:

1. *Squaring:* This is the process of squaring the recorded dimensions and entering the resultant lengths, areas and volumes in the fourth or squaring column (D) of the dimension paper.
2. *Abstracting:* The transferring of the squared dimensions onto an abstract paper in a recognised order under the appropriate work sections. The figures are subsequently totalled and adjustments made for deductions.
3. *Billing:* The transferring of the abstracted items onto a bill paper set out under the appropriate work section headings and in a recognised order. Descriptions are given in full at this stage and the quantities are given in a recognised unit of measurement (i.e. m^3, m^2, No/Nr and tonne).
4. *Editing:* The editing of the draft bill is usually carried out by a partner or a senior member of staff to ensure conformity with the accepted standard and a firm's bill production practice.

5. *Typing:* The typing of the draft bills of quantities may be either done in-house or sent out to a company offering typing, printing and binding services.
6. *Reading over:* The reading over of the typed draft bill. In the process, all typing errors are corrected and corrections are rechecked before printing.
7. *Printing:* Printing and binding the bills of quantities may be done in-house (where the quantity surveying firm has the facilities) or by a company of printers.

7.3.2 Direct billing

The traditional working-up process, which has been used by quantity surveyors for many years, is a lengthy and tedious operation. Direct billing is one of the ways that has been devised to shorten the working-up phase of the bills of quantities production. This method eliminates the abstracting function by transferring the taking-off items direct from the dimension paper onto the bill paper in a recognised order. The direct-billing method is suitable for a simpler project or a work section with a limited number of work items.

7.3.3 Modern methods of bill production

Although it has been branded as painstaking and time consuming, the traditional system of taking off and working up, which involves abstracting and billing, has been in general use for many years. However, in recent years, the development of more efficient systems has reduced or eliminated the working-up processes. The most important techniques in current use are direct billing, cut and shuffle and computerised bill production.

Cut and shuffle

Cut and shuffle is an expedited method of producing bills of quantities using specially designed dimension paper. Each cut and shuffle paper is ruled into four separate slips (see Figure 7.2) and during the taking-off process only one item is entered on each slip. Generally, two copies of dimension paper are required; the taker off retains the original and, after squaring, casting and checking, the copies are cut into slips, collected and sorted into bill order without the necessity for the abstracting and billing operations. As the system alleviates the need to write and check the abstract and the draft bills of quantities, full descriptions are given to all measured items irrespective of their location in a building, and no abbreviations are used. Although this approach prolongs the taking-off time, the time taken to produce the finished bills of quantities is shorter than when the traditional method is adopted.

The advantages and disadvantages of cut and shuffle may be summarised as follows:

F				F				F			
G				G				G			
			H				H				H
B	C	D	E	B	C	D	E	B	C	D	E
		J	K			J	K			J	K

B = Timesing column – where multiplying figures are set against description of an item of work that occurs several times within the measurement.

C = Dimension column – where the sets of dimensions relating to the description of items of work are entered. These may be lineal, square cubic, number or kilograms.

D = Squaring column – where sets of dimensions are multiplied out.

E = Calculation column – where preliminary calculations and location notes are entered.

F = Work section reference column – where work section and any additional reference numbers are entered.

G = Description column – where the description of items entered and written in full.

H = Totals of reduced dimension column – where the total of the squared, collected and reduced dimension is entered.

J = Dimension totals column – where the total of the squared dimension is entered.

K = Slip number column – where the slip number is entered.

Fig. 7.2 Sample of cut and shuffle paper

Advantages

- There is no repetitive writing of item descriptions, and this reduces time taken to produce bills of quantities.
- The laborious tasks of preparing and checking abstracting and billing are eliminated.
- The system lends itself to greater use of clerical staff in lieu of highly paid technical staff in bills of quantities production.
- It is a semi-mechanical method and, hence, lends itself to greater use of office machines.

- The coding which this system applies facilitates the introduction of computer techniques to bills of quantities production.

Disadvantages

- The taking-off phase of the bills of quantities production is prolonged because of the need to write out a full description for each bill item.
- The code or reference required for each slip can be time consuming and a wrong code/reference can cause complications which may take some time to sort out.
- The system does not lend itself to the training of junior staff in the production of bills of quantities using the traditional approach.
- Searching for bills of quantities items in the original dimensions for final account work can be time consuming.

Computerised bills of quantities production

Computerised bill production is the use of computers to aid the process of bills of quantities production. This method is gaining popularity, primarily because of the speed at which bills of quantities can be completed after the initial measurement of the work sections, and even currently there are some computer packages which can computate quantities direct from a computer-aided design (CAD) systems. Generally, computerised bill production requires a comprehensive standard library of descriptions based on the current standard method of measurement for building works. Standard libraries of descriptions are prepared and coded, then transferred to magnetic tape to form the computer's memory bank or store.

The taking-off phase for computerised bills of quantities does not vary much from that of the traditional approach, except that a different dimension paper is used. This dimension paper has the normal four columns for dimensions and descriptions on its left side but has also a separate column for coding. After the normal manual taking off, the coded information is then fed to the computer which executes the squaring, reducing, billing and printing. Generally, it is a quick approach to bills of quantities production which alleviates manual squaring, abstracting and billing processes.

7.4 FORM OF BILLS OF QUANTITIES

All bills of quantities, irrespective of format, should contain a complete description of the works (i.e. quality and quantity expected) and the conditions under which works are to be executed. The basic information found in a typical bills of quantities is invariably arranged under the following headings:

7.4.1 Preliminaries

The first section of bills of quantities is normally termed *Preliminaries Bill*. This section of the bills of quantities describes the nature and extent of the work, the type of contract and all factors affecting the physical execution of the works and, hence, contains several non-measurable important financial matters which relate to the construction project as a whole. These financial matters are not confined to any particular section of the project and, at tender stage, the contractor is given the opportunity to price them accordingly.

Most preliminary items are contained in the standard method of measurement of building works (SMM) and matters covered in the preliminaries bill may include the following:

- Names of parties to contract.
- Description of works.
- Drawings and other documents.
- Parties and consultants.
- Forms and conditions of contract with clause headings.
- Contractor's liability.
- Employer's liability.
- Obligations and restrictions imposed by the employer.
- Contractual requirements for insurances and to site facilities.

The following items where required, may also appear in the preliminaries section of the bills of quantities.

- General facilities and obligations such as plant, tools and vehicles, scaffolding, site administration and security.
- Transport for work people.
- Health and safety of site administrators, operatives and visitors.
- Protecting the works from adverse weather and vandalism.
- Water, lighting and power for the works.
- Temporary roads, site accommodation and telephones.

7.4.2 General preambles

This is the second section of the bills of quantities which describes the quality of materials, processes and workmanship required to complete a construction project. The general preambles usually provide the contractor with the following information:

- Type and quality of materials, equipment and fixtures including the relevant British Standard (BS) references.
- Quality of workmanship and permissible tolerances.
- Methods of fabrication, installation and execution.
- Test and code requirements.
- Gauges of manufacturer's equipment.

Unless the general preambles are read in conjunction with the information shown on the drawings and the measured works section of the bills of quantities, the contractor will be unable to price adequately the items contained in the measured section of the bills of quantities. To promote quick reference, descriptions of items in the measured works section of the bills of quantities are made up to bear the same preambles reference numbers. By this arrangement the quantity surveyor is able to shorten the description of the measured items and also to expedite cross-referencing during the project's production phase.

7.4.3 Measured works

The measured works section of the bills of quantities contains the schedule of quantified and fully described items of work. Depending on the amount of construction information available at the bill preparation stage, work items contained in this section of the bills of quantities may be measured as either firm or approximate. For this reason, the bills of quantities may be classified as follows:

1. *Bills of firm quantities:* These are bills of quantities which, except for a few items, contain accurate quantified items of work. Any item of work which cannot be measured firmly due to lack of information is qualified with the words *'provisional or approximate quantities'* and is subject to later remeasurement and adjustment.
2. *Bills of approximate quantities:* These are bills of quantities which contain work items which, as a result of insufficient information, have been quantified approximate only. However, the descriptions of items of work should be accurate and only the quantities are subject to later remeasurement and adjustment.

The following items will also appear in the measured works section of the bills of quantities.

Prime cost (PC) sum

Sections of work specifically intended to be executed by specialist firms are covered by prime cost sums in the bills of quantities. Prime cost (PC) sums represent various sums of money written into the bills of quantities either for works which are required to be carried out by a nominated sub-contractor or goods and materials required to be obtained from a nominated supplier. To enable the client to receive or accept a tender that represents the total cost of all the works, the contractor is obliged to include these sums in the tender and, hence, is given the opportunity to price for profit and attendance. When the prime cost sum is adjusted, the profit amounts are also adjusted pro-rata by applying the same percentages (i.e. those inserted in the bills of quantities against the PC sums at tender stage) to the actual costs. Specialist work varies but may include piling, structural steelwork, electrical and mechanical installations.

Furthermore, the attendance contractors provide to specialist sub-contractors during the execution of the specialist work is classified as either general or special. General attendance means granting a specialist sub-contractor the use of a contractor's site facilities such as standing scaffolding, messroom, sanitary accommodation and welfare facilities, space for office accommodation and for storage of plant and materials, light and water. Special attendance, on the other hand, refers to attendance provided to a specialist sub-contractor over and above the general attendance provisions, and this includes cost items such as access road, hardstanding, storage, power, unloading, distributing, hoisting and placing materials/components in position.

Provisional sum

A provisional sum is a pre-assessed sum included in the bills of quantities by the client to cover items of work or for costs which cannot be defined or detailed owing to lack of information or uncompleted design. In order to eradicate the confusion and the disagreement that the inclusion of provisional sums in the bills of quantities creates, provisional sums are currently classified as either undefined or defined. The reasons for this classification may be summarised as follows:

1. *Undefined provisional sums:* Normally the contractor is not expected to make any allowance in the contract programme, plan or price in the preliminaries for items of work covered by undefined provisional sums. Therefore, the contractor is reimbursed from the provisional sum for any extra costs he or she incurs on all items that come under the description of undefined provisional sums.
2. *Defined provisional sums:* The contractor is required to make an allowance in his or her contract programme, plan and price in the preliminaries for items of work covered by defined provisional sums. Therefore, during the progress of the works, no claims for reimbursement are entertained. To enable the contractor to provide adequately for items of work described under the defined provisional sums, the following information are included in the bills of quantities.

   (a) The nature, extent, quantity and quality of the work.
   (b) The location of work, how it is executed, limitations and whether subsequently covered.

Daywork

During the progress of the works, any varied work which, by its very nature, cannot be properly measured and valued, is calculated at daywork rates based on prices prevailing at the time of work. The normal practice is that a provisional sum (probably undefined) is included in the bills of quantities to cover any dayworks. The intention of such provision is to provide a standard basis of evaluating work which can only be assessed on a prime cost basis. Therefore, at the tender stage, the contractor is asked to insert a percentage addition required on labour, material and plant to cover overheads and profit together with other costs not included within the prime costs (e.g. weather uncertainties, bonus).

7.5 CLASSIFICATION OF BILLS OF QUANTITIES

Bills of quantities can be produced in several formats; however, the following have been selected for consideration.

7.5.1 Trade order or traditional bills

Trade order bills of quantities form a contract document which has been presented in trade sections and sub-sections, following the sequence scheduled in the current standard method of measurement for building works. Bills of quantities of this type are also known as *Traditional bills* and provide a good basis for competitive tendering but, the project information they contain cannot easily aid the organisation of production and the management of cost.

The advantages and disadvantages of traditional bills of quantities may be summarised as follows:

Advantages

- They group total quantities of both labour and materials together in one trade and, hence, facilitates the assessment of the total cost of each trade within a project.
- As they have been in use for many years, users have become accustomed to the general layout and, moreover, the bills facilitate the production of estimates/tenders.
- Traditional bills of quantities assist contractors in programming the works in terms of what trade to bring on to site, when and for what duration.
- They form a valuable document which promotes the adoption of the *NEDO* (*National Economic Development Office*) *Price Adjustment Formula* for the adjustment of price fluctuations of work items contained.
- They form a very useful document for contractors when ordering materials for each trade as he or she can check quantities with the bills and reconcile any differences.
- All measured items contained can be easily traced when adjustment is required during valuation of variations and/or preparation of the final account.

Disadvantages

- Traditional bills of quantities give no indication of the degree of repetition of some items of work in different locations throughout the construction project (e.g. concrete floor at all levels in a building are measured as one item).
- There is no induction of the location of items of work or of the order in which works are to be executed to enable contractors to price them accordingly and adequately.
- They do not assist contractors in the assessment of materials or labour required for any specific part or element of the building at any given time.
- Quantity surveyors cannot use them for the preparation of elemental cost planning unless a laborious and time-consuming elemental cost analysis exercise had been carried out beforehand.

- The use of traditional bills of quantities makes preparation of interim valuations or the assessment of value of variations a difficult operation.

7.5.2 Elemental bills

Elemental bills of quantities are arranged or divided into sections, each of which is an important element of a building (e.g. external walls, roof, floors) as opposed to trades. Within each element, the work items may be billed in trade order. By adopting this format, the location and cost of an element are readily identifiable in the bills of quantities and, hence, assists the elemental cost planning of future construction projects. However, the appearance of an item several times in the various elements in the bills of quantities makes the tendering process laborious and more complicated.

The advantages and disadvantages of elemental bills of quantities may be summarised as follows:

Advantages

- They indicate the location and order in which the work is to be executed.
- They assist contractors in ordering materials or labour for any specific part of the building at a given time.
- On receipt of tenders, they assist quantity surveyors in the preparation of elemental cost analysis and future cost planning.
- They differentiate between similar items of work in different locations throughout the building.
- They facilitate the preparation of interim valuations, measurement and valuation of variations.

Disadvantages

- They group together labour and material requirements within each element and estimators find it difficult to collect and assess total quantities of materials or type of plant required for the project.
- Time taken for the production of the elemental bills is longer than that for trade bills due to repetition and the number of times work items are written into the bills of quantities.
- As a result of the repetition and, hence, the increased number of items required to be priced, the production of competitive tenders by this method takes longer than when trade bills of quantities are used.
- These bills are not so useful for the adoption of the *NEDO Price Adjustment Formula* method in price adjustment fluctuations.

7.5.3 Operational bills

The operational bills of quantities is a form of contract document which divides work into actual site operations or stages of the work as distinct from either trades or elements. An operation, in this instance, is regarded as the amount of work that can be produced by a gang of operatives at some definite stage in the construction process without any interruptions. Under each operation, a description of the work to be executed is given; the quantities of materials required and, where applicable, plant requirements for each operation are also set down separately. However, the contractor decides on the amount of labour required to complete an operation. In effect, an operational bill describes the amount of work to be completed during the particular operation and this provides a complete breakdown of the costs which then facilitates production organisation and project cost control.

The advantages and disadvantages of operational bills of quantities may be summarised as follows:

Advantages

- They assist contractors in the preparation of a programme for the construction works.
- They expedite identification of high cost operations (when tenders are received) and the introduction of measures to effect cost savings.
- They facilitate the preparation of a critical path analysis.
- The schedule of materials and the operational sequence taken together assist contractors in ordering materials and labour for any particular operation.
- They differentiate between similar items of work at different floor levels or locations, and this enables contractors to submit a tender which reflects the variations in the difficulty of the construction works.
- They enable contractors to keep accurate records of the actual costs of constructing each part of the construction project.

Disadvantages

- Operational bills consume more time to prepare and also to price; and because of this, are now only used for very exceptional projects.
- Contractors neither accept nor approve of a tender document that dictates to them how they ought to execute construction work.
- Variations under operational bills are extremely difficult to value.
- Successful tenderers may not wish to carry out work in the sequence proposed in the tender documents.

7.5.4 Sectionalised trade bill

Sectionalised trade bills are basically *trade* bills of quantities with each trade sectionalised into elements. This type of bill has the advantages of both *trade* for tendering and *elemental* for construction production management. The advantages and disadvantages of sectionalised trade bills are the same as that for elemental bills of quantities.

7.5.5 Activity or locational bills

Activity or locational bills refer to operational bills of quantities measured in accordance with the standard method of measurement for building works but in which the items are billed in sections which relate to activities established by network diagrams. For this reason, the items are coded in the bills to correspond with their general location in the network. Activity bills of quantities have the advantage of facilitating the identification of production delays and their effect on the production programme. Nevertheless, the preparation of activity bills requires skill in the production of accurate network diagrams.

7.5.6 Annotated bills

These types of bills of quantities may be in any of the formats discussed above, but side notes are incorporated to indicate the location of the measured work items within the construction project. The location notes are helpful to the estimator at the tender stage and to the construction team also at the production stage.

The advantages and disadvantages of annotated bills of quantities may be summarised as follows:

Advantages

- They assist contractors with respect to the identification of location of work items and, hence, the contractor is able to estimate the cost of the works accurately.
- They facilitate the measurement of variations and identification of mistakes in the bills of quantities by the contractor and the client's quantity surveyors.

Disadvantages

- The locational notes provided in the document consumes more time in its preparation and, therefore, annotated bills of quantities cost more to produce.

Generally

Despite their wide acceptability and usage, increasing scepticism has arisen recently from some clients and other construction professionals about the benefits of the continued use of bills of quantities. These sceptics claim that bills of quantities have run their course and that buildings can and must be produced without them. This belief, and the general desire for change, have increased a search for alternative ways of construction procurement. Those who argue strongly against the use of bills of quantities contend that:

- The production of bills of quantities is a time-consuming and costly exercise which the client can not only do without but can also save money on professional fees.
- Their production increases the construction procurement time leading to a loss of revenue or by the client's usage of completed facility.
- The usefulness of bills of quantities is limited as construction professionals cannot guarantee their accuracy in descriptions and quantities; this, therefore, presents an unreasonable risk to users of this type of documentation.

As the motives of advocates and opponents (of the use of bills of quantities) are not the same, one expects that the debate on the use of bills of quantities will continue into the foreseeable future until a better and acceptable construction documentation and procurement method is found.

7.6 ADEQUACY OF CONTRACT DOCUMENTATION

Many professionals and clients in the construction industry assume that, due to the complex nature of construction projects, construction disputes are inevitable. Moreover, the dispute-prone construction industry's position has been exacerbated by the separation of the design and production functions, increased number of specialist participants and the resultant production and utilisation of increased quantity of construction information. In addition, the general practice of calling for bids on an inadequate contract documentation is one of the prime sources of construction disputes. This, therefore, places the client's professional advisers under great pressure to improve the quality of contract documentation in the face of adverse factors, namely planning delays and client's uncoordinated efforts.

7.6.1 Planning delays

Delays experienced in obtaining planning approval have perhaps the most disruptive effects on the construction process. Planners frequently control the pace of the construction process at the initial stages of the design phase. Hence, delays to design due to approval procedures are endemic and these can have a serious effect on the overall project's programme. In addition, great uncertainty exists for the client as to whether or not the requisite planning permits will eventually be granted. All these delays and uncertainties, eventually, curtail the time required for the preparation of adequate contract documentation.

7.6.2 Client's uncoordinated effort

Client's uncoordinated action which contributes to the production of inadequate contract documentation are numerous and may include:

- A poorly defined brief which is lacking in content and, hence, provides incomplete information on the client's requirements.
- Failure of clients to plan project's requirements thoroughly and well in advance, leading to the creation of unnecessary variations during the design phase.
- Initiating project late and pressing the design team to get on with the job of producing contract documentation when the project design is nowhere near completion.

To ensure their positive design input, design teams do advise clients of the need for a clearly defined brief, advance planning of their needs and early planning application. Construction clients are also advised to spend more money at the front end of their construction projects and give the design teams more time to plan and design the project in a more meaningful manner before calling for competitive bids.

7.7 GENERAL

Construction disputes are always about unanticipated extra costs or variations and the only way to reduce significantly the opportunities for cost overruns is through preparation of adequate and more comprehensive contract documentation. Therefore, although all construction projects are not the same, where applicable a client would expect that his or her professional advisers undertake the appropriate measures to ensure the production of adequate contract documentation. Hence, the client's professional team may adopt the following measures to improve the adequacy of contract documentation.

- Drawing the client's attention to the fact that a good design is an investment and, like all investments, it requires adequate funding.
- Being aware of the need of contract documentation to clearly identify, describe and, where appropriate, correctly quantify that which is to be estimated.
- Endeavouring to make all drawn and written information complement each other clearly so as to avoid the danger of breakdown in communication.
- Drawing client's attention to the impracticabilities of the allocation of a short construction project time and of carrying out a rush job which goes out to tender on a far from complete design.
- The involvement of a quantity surveyor at the initial stages of the design process to contribute to the interchange and/or the introduction of design ideas based on elemental cost information.
- The promotion of efficient teamwork by the architect such as an early discussion with all design team members and the establishment of formal communication channels at the critical design stage to ensure free flow of project information.
- Giving sufficient thought and time to ensure that all the design team members fully understand the client's project objectives at the initial stages of the design process to effect expedient and satisfactory design solution.
- The provision of manufacturers'/suppliers' names and addresses for proprietary components in the contract documentation.
- The inclusion of information on client's imposed restrictions and risky/difficult construction works in the contract documentation to enable tenderers to price adequately.

The aim of all the above measures is to ensure that the contract documentation for the client's project construction contains all the relevant information which will enable it to be executed free of disputes and expensive contractual claims.

SUMMARY

In this chapter we have wrapped up the construction contract documentation and its respective functions. The production of contract documents are important as they enable the client's requests to be established before contract. Furthermore, they act as a benchmark against which the quality of production is to be measured. Additionally, they establish the basis for payment to the building contractor. The major documents which enable prices to be obtained before contract are the bills of quantities, although their continued use is now being questioned as some major projects have been undertaken successfully in the UK without their use. However, as public sector clients favour the use of bills of quantities in spite of this criticism, it is likely that they will continue to be used for many years hence.

FURTHER READING

CIOB 1992 Inadequate tender documents and procedures; who pays? *CIOB Discussion Document.*

Harrison, H.W.; Keeble, E.J. 1983 Performance specification for whole building. *Report on BRE Studies, 1974-1982.* BRE.

Ramus, J.W. 1993 *Contract practice for quantity surveyor.* B.H. Newness.

Seeley, I.H. 1975 *Building quantities explained.* Macmillan.

Turner, D.F. 1983 *Quantity surveying practice and administration.* George Godwin.

Willis, C.J.; Ashworth, A. 1990 *Practice and procedure for the quantity surveyor.* BSP Professional Books.

CHAPTER 8

ESTABLISHMENT OF PRODUCTION PRICE

8.1 INTRODUCTION

At one point in the construction development process, there is a need for the establishment of a production price; hence, prices are normally obtained from building contractors who are willing to undertake the production of the construction project. The techniques adopted in the process of contractor selection and price determination are complex. It requires both intelligence and pre-qualification phases in which the proposed contractor's past business record, current operational performance and financial standing are vetted. At the end of the pre-qualification process, the most suitable contractors are listed for bidding. This is followed by an invitation to competitive tendering, tender evaluation and so on.

8.2 PRODUCTION COST

Production price represents the cost that the client incurs in having his or her construction project built. This cost is composed of builder's site production costs, plus overheads and profit, and in the construction industry this cost is normally established before commencement of site production. The production price may be established either by negotiation or by competitive tender. When negotiation is the client's preferred contractor selection option, a contractor with expertise in the project is approached and the production price is negotiated. Competitive tendering, on the other hand, is a contractor's selection procedure in which a number of suitable contractors are invited to submit competitive bids for the execution of the project. Therefore, at some stage in the client's project development, there is a need to identify suitable contractors to be invited to submit prices for the execution of the construction project.

8.3 CONTRACTOR SELECTION PROCESS

Selection of contractor for the construction of a project is one of the crucial decisions in the client's development ambitions. Ill-conceived decisions regarding this will impair the client's prospective development gains. The criteria for contractor selection may be price, time or both in addition to the contractor's expertise in a client's development project. If price is the selection criterion, a contractor's tender represents the price that is offered to execute the works in accordance with the contract documentation. This is an objective issue as the client seeks the most economic price for the development. However, time and expertise criteria are subjective issues as because of the need for expedited construction programme and good quality workmanship, a client may seek a price from a sole contractor.

Generally, a good contractor for a project is an important factor contributing to its successful completion and the client's professional advisers may look for the following attributes when drawing up a list of suitable contractors from whom to invite bids.

1. *Contractor's reputation in business:* A contractor's good past record of performance in the construction business is considered; for example, the number and monetary value of projects the contractor has completed successfully and high standard of workmanship are the attributes of a reputable construction company.
2. *Contractor's financial standing:* The knowledge of a contractor's current financial standing or financial record enables the client to assess whether or not the company is likely to go into liquidation and, hence, there is the need for a bank guarantee or caution when dealing with that contractor in business.
3. *Contractor's potential resources:* The contractor's potential resources embrace his or her physical as well as human resources. For example, buildings, offices, workshops, factories, fixed plant and machinery; the number and type of trade operatives regularly employed and the quality of management personnel (at the head office and on site), their technical knowledge and experience.
4. *Contractor's normal conduct of business:* Considerations under this heading include the type of work package normally undertaken by the contractor's own direct operatives and those that are normally placed as sublets; the categories of clients (e.g. public, private) the contactor chooses to work for and the type of projects in which the contractor specialises (e.g. new build, maintenance or refurbishment).
5. *Contractor's attitude on contractual claims:* Contractors who have a good record of not being claim conscious or claim loving and who take legal action only as a last resort is a sign of flexibility in resolution of construction disputes.
6. *Non-economic factors:* Contractor may be selected for reasons unrelated to performance; for instance, a contractor may be appointed to foster business relationship (subsidiary company) or to maintain or promote employment in a locality.

8.4 TIMELY SELECTION

In order to achieve a client's development objectives, his or her professional advisers normally determine the most advantageous moment to select a building contractor for the construction project. Their concern will either be selecting the contractor before or after the design phase. This decision is crucial and in most cases is influenced by the client's development strategy and many other factors, including:

- Type of development – standard system building, a building of specific design or refurbishment of an existing facility.
- Need for prompt start on site.
- Shorter project time requirement.
- Complexity of development and a need for early contractor involvement.
- Certainty of financial commitment before contract.
- Degree of financial risks client is willing to assume.
- Need for strict cost control.
- Scale of development and number of participants.
- Availability of resources.
- Degree of unquantifiable risk involved.
- Business relationships and client's satisfaction.

Generally, for projects characterised by complexity, large scale and shorter completion time, contractor selection during the design phase is a better option. In this approach, the contractor is able to advise on a buildable design solution, selection and availability of resources and project planning. However, on projects where the client requires certainty of total financial commitment before contract and strict cost control, contractor selection *after* the completion of design and contract documentation will better suit the client's requirements.

8.5 CONTRACTOR SELECTION METHOD

The methods of contractor selection can be described as either by competition or by negotiation. In either case, the decision taken should reflect the client's development aims – i.e. the completion of his or her construction project economically, safely, quickly to the required quality and at a profit. In construction projects, the most recognised and effective method of contractor selection is known as tendering, which may be described as an offer in writing to execute defined work under stated conditions at a price. Tendering procedures cover the various methods that are used by the client's professional advisers to obtain offers to execute the construction project at a price. The offers are made by the contractors approached, who in turn submit prices based on assessment of their production cost and margin level to cover their overhead and profit. The procedures adopted by contractors in the tendering process do vary, but each approach is aimed at securing yet another job and, hence, survival. The type of tendering arrangements available to clients may be grouped under the headings of competitive tendering and negotiated tendering.

8.5.1 Competitive tendering

The competitive tendering concept is deeply rooted in the UK construction industry. The basic idea behind this concept was that competitive tendering protected the public from extravagance, corruption and other improper practice by public officials. The original function of the competitive bidding requirement was to ensure that the public received the full benefit of the free enterprise system by letting building contractors provide clients with services at the lowest price offered by competitive tendering. The competitive tendering system has two major advantages. It assumes that the lowest bidder is the construction company that will be able to make effective use of its resources and manage the project to the benefit of the client. It also assumes that the bidding process will be independent of any form of pressure (such as political, social, economic). Its objectivity is ensured when price is the sole criterion for evaluating the tenders. However, the system has numerous disadvantages, the major one being that the selection process is based only on one element – cost. Other elements, such as quality of workmanship and time are not accounted for. The system is also fraught with unreasonably low bids, bid rigging, cartel and so on. The types of competitive tendering arrangements practised may be grouped as follows:

Open tendering

Under this method, an advertisement is placed in local, national and technical press by the client's professional advisers inviting interested building contractors to submit tenders for the client's proposed construction project. In the simplest terms, any contractor who is legally trading may request to tender; however, a deposit is paid for the contract documents, which is refunded upon receipt of a bona fide tender.

The risk element in this method of tendering to a client is high as he or she cannot ensure before the bidding that the prospective building contractor has the required experience, a sound current trading position and a sound cash flow situation. However, tender documents are despatched to contractors who indicate their willingness to submit tenders for the proposed project.

Public authorities sometimes invite open tenders and it is a method most suitable for small building and maintenance works as well as unusual specialised projects such as asbestos removal and maintenance work to swimming pools.

The advantages and disadvantages of open tendering are summarised as follows:

Advantages

- Open competitive tenders make it possible for keen contractors to obtain the work and offer a good service to clients.
- It helps contractors willing to grow or expand their market segment to find new clients.
- As the public sector clients seek prices through this method, it provides work for local construction companies who rely on their area public sector client for work and hence, respond positively to the press advertisements.

Disadvantages

- Increased cost of tendering in terms of extensive production of contract documentation and estimating.
- Less satisfactory contractor performance may leave the client with a bad job which may be expensive to rectify.
- There is no guarantee that all tenders received are bona fide.

Single-stage selective tendering

Under this method, tenders are either invited from a current standing list of approved contractors or from a short list of contractors who have responded to an advertisement in the national and technical press (which has set out details of the client's proposed construction project).

The Code of Procedure for Single Stage Selective Tendering 1991 sets out guiding rules for this system and it also makes the following useful recommendations.

- Depending on the size of the construction project, five to eight contractors should be invited.
- Bidding contractors should be given adequate notice and details of the proposed construction project to enable them to ascertain whether a spare capacity exist in their establishments to undertake the proposed work.
- The contract period should be specified in the tender documents to restrict competition to price alone. However, separate offers based on a revised time are not to be discouraged.
- Tender period should be about four weeks; nevertheless, this may be varied to reflect the size and complexity of the proposed construction project.

The advantages and disadvantages of single-stage selective tendering are summarised as follows:

Advantages

- Client obtains a cost-effective price which reflects the state of the national economy and what the construction market can bear at the time.
- Contractors submit a bid with clear awareness that when successful there is spare capacity in their establishments to accommodate the project.
- It satisfies public accountability requirements in construction projects for the public sector clients who are expected to fulfil this requirement.
- Client is not responsible for contractor's programme of work and site operations and therefore assumes less financial risks.
- Tight cost control technique can be applied (during both design and production phases) to ensure that a client's development budget is not exceeded.
- Client can ascertain his or her total financial commitment before contract and is therefore able to plan and arrange for it.
- More time is allowed for project planning and design and hence the client receives not only a good buildable design but, on completion, a good product.

- It results in less abortive tenders and, hence, a reduction of tendering cost and waste in the construction industry.

Disadvantages

- The short-list principle may exclude suitable building contractors who are capable of providing clients with a good service.
- Contractors are excluded from design decisions, and their expertise on buildable designs, selection of suitable materials and production methods is not fully utilised.
- Useful project time is lost in detailed planning and preparation of scheme before invitation of tenders.
- Contractors capitalise on variations issued by the client during production and this leads to increased cost of production and construction disputes.

Two-stage selective tendering

The two-stage selective tendering system involves competitive selection of building contractors in the first stage and negotiation in the second. Competitive selection in the first stage is based on pricing approximate or notional bills of quantities containing work similar to the project in hand or, when available, pricing bills of quantities on a similar or related project.

The Code of Procedure for Two Stage Selective Tendering 1983 recommends a maximum of six tenderers irrespective of the size of project and a minimum tender period of five weeks. The code further recommends the following:

- Bidding contractors are required to be given adequate notice and details of the proposed construction project to enable them to decide on the willingness and ability to tender.
- Acceptance and consideration of an additional alternative offer which varies any aspect of the project supported by fully specified and priced build-up.
- Rejection of all qualified tenders.
- Inclusion of specified contract period in tender documents to restrict competition to price only.

When a contractor is selected after the first stage, he/she is integrated into the design team. As a construction professional the contractor plays an active role in providing design advice on quality, buildability, programming and cost. In the second stage, the total contract price is determined partly by negotiation and partly by pricing based on data provided by the contractor at the first stage. Under this method, the element of financial risk assumed by the client is greater than the single-stage selective tendering approach. The factors that may influence contractor selection under this method include the following:

- Experience, technical knowledge and ability to execute the proposed construction project.
- Capacity, in terms of physical resources (workmanship, plant stores, etc.) and human resources (quality of management, design staff and operatives).

- Reputation based on performance on past construction projects, speed of construction, good quality of workmanship and after-contract service.
- The ability to undertake research and development aimed at solving problems posed by the construction project or before they arise.
- Length of time in business, sound current financial and trading position.

Two-stage selective tendering is generally suitable for clients whose development programme requirement does not allow sufficient time to complete design before contractor selection. The circumstances that may bring this situation about include construction projects where:

- The benefits to be accrued from early start and shorter construction time exceeds the likely risks of commencing the work on half-completed design information.
- Early contractor involvement to advise on buildable design, programming and coordination is required.
- Instant start is necessary before the end of the financial year leading to savings in taxation or benefits of government grants.
- Separation of design from construction is impractical (e.g. standard or system building where the contractor can make a considerable technical contribution to the design).
- Price is only one of the criteria for selection and where design input from the contractor is required.
- Client wishes to assume a greater proportion of the financial risks.
- Specialist sub-contract works form the bulk of the total construction cost and, hence, require the input of specialist sub-contractors, their early integration and coordination by the main contractor.
- There is an extreme shortage of materials and advance purchases of essential materials and components is beneficial to the client.

The advantages and disadvantages of two-stage selective tendering are summarised as follows:

Advantages

- The client benefits from the designer/contractor collaboration during the design phase leading to a buildable design.
- The client benefits from both elements of competition and negotiation in contractor selection.
- Early involvement enables the contractor to organise his or her resources and plan for the construction phase more successfully.
- Early start on site and completion can be more easily planned to meet the client's development strategy.
- Design and construction phases may be overlapped to shorten the duration of the construction process.
- Contractual claims are minimised as contentious design details are identified and modified before contract and/or site production.

- Advance purchasing of essential plant, materials and components can be planned and effected to ensure their availability when required.

Disadvantages

- During the negotiation stage, the contractor may be able to alter his or her position/price when the full effect and complexity of the construction project is known.
- More effort is required from the client and his or her professional advisers than for single-stage selective tendering due to price determination on incomplete design information.
- Legal responsibilities for design failures are not easily apportioned due to the multiple design input.
- The contractor may impose his or her favourite design solution and detailing on the design team.

8.5.2 Negotiated tendering

Under this method of contractor selection, the client approaches the contractor of his or her own choice with a view to this being the only firm who will submit a tender. The contractor selection is based on reputation, specialised skills, sound financial position as well as business relationship.

The process of negotiation can be time consuming, requiring the skills and energies of experienced negotiators. To ensure fairness and the ability to negotiate effectively, the parties to the negotiations should be of equal or equivalent positions in their respective organisations and also should possess the same information on which the basis of negotiations will be established. The negotiation can be based on either nominated bills of quantities or cost assessment from the first principle of estimating.

Nominated bills of quantities

Under this approach, the contractor is obliged to provide a *nominated bill* which is a copy of priced bills of quantities for a contract that the contractor has won in competition. From the nominated bill the following exercises are carried out:

- The assessment of any major differences between the *nominated bill* contract and the new contract is made. Onerous contract conditions, the differences in the tender levels, the state of the construction market, the availability and cost of resources, labour, plant, materials and finance are taken into account.
- The contractor is asked to analyse the unit rates of major items of work to facilitate identification of elements of establishment charges and profit.
- The contractor is asked for a breakdown of preliminaries in order to establish the cost of major items of plant and site management charges.

Based on the above information, new rates can be negotiated to reflect all known market factor changes and financial risks that may affect either the client or the contractor in the execution of the proposed construction project.

First principle of estimating approach

This method is used when a suitable *nominated bill* cannot be provided owing to the magnitude and complexity of the client's proposed construction project. In such a circumstance, the contractor's establishment charges and profit element, cost on major items of plant, site set up and management costs are negotiated. Once agreement on these factors has been reached, unit rates are calculated for individual items in the bills of quantities from first principles of estimating. The process entails utilising suppliers' prices of materials and components, all-in rates for labour, plant hire rates and sub-contractor's prices.

It must also be said that, during the process of negotiations, both parties do not necessarily bind themselves immediately to any of the costs that emerge. The negotiations thus proceed through proposals and counter-proposals of various rates and calculations until finalised. At this point the magnitude of the project in terms of cost and the financial risks are known; and the position of eventual agreement reached, leading to a legally binding contract for both parties.

Negotiated tenders may be adopted in construction projects where:

- Magnitude of the project cannot be assessed before production (e.g. due to its complexity).
- Immediate start is urgently required (e.g. emergency work).
- Selected contractor can offer a specialised service to a client because of special plant, finance, experience in similar works, awareness of client construction problems or past satisfactory business association.
- Contractor's early involvement is urgently required to assist the design team with advice on buildability, material selection and the like.
- Retention of the contractor in the execution of the client's programme of construction projects.
- Contractor and client may have a special business relationship making negotiation a beneficial option for their respective businesses.
- Very few contractors are available with the skill and experience required for the particular type of project.

While the system may be usefully employed in the above situations, it should be mentioned that negotiated tenders may be very difficult to conclude and may also require the efforts of experienced staff from both the client professional advisers and contractor's establishments.

The advantages and disadvantages of negotiated tenders are summarised as follows:

Advantages

- Early contractor appointment in the design phase leads to a beneficial contribution as his or her skills and work experience are made available to the design team.
- Early start on site may be achieved to meet the client's development strategy.
- Client's tendering costs are substantially reduced owing to the production of minimal tendering information.

- If overlapping of design and construction phases eventuates as a result of the negotiation, it condenses the development period with attendant savings in costs.
- It minimises tardy delivery of projects as the contractor can carry out some construction planning during the progress of negotiations.
- Client obtains the contractor he or she prefers as the contractor is selected for ability as well as price.
- All the important points of the construction project (e.g. construction programme, method and procedure) are discussed during the negotiation and this effects the rational price.

Disadvantages

- Client obtains an offer which is not truly competitive and does not reflect what the construction market can bear.
- There may exist legal implications of joint design when the negotiation results in design and production overlap.
- Contractor experiences difficulty in estimating on outline information.
- It may not satisfy the requirements of public accountability in construction projects for public sector clients.

Each of the above contractor selection methods has its merits and demerits. No one method is necessarily better or worse than another and, therefore, each must be viewed in the light of the circumstances appertaining to the client's development strategy.

8.6 ESTABLISHMENT OF PRODUCTION PRICE

The establishment of production price takes place under a process known as *estimating and adjudication*. It is a process in which contractors assess a construction project's cost after identification of the most competitive base cost of each and every element of a tender and the assessment of the potential commercial risk and reward factors relating to them. Moreover, it is a yardstick of performance interpreted into cost terms of labour, plant and material. It must therefore be factual and systematic and closely related to the way in which the project funds will be spent. When performing this important task, building contractors apply their estimating skills, adopt different strategies, require varied information and use diverse decision-making processes to compile the most economically advantageous and competitive bid. The information required by contractors for the cost establishment process are many. While most of them are provided by the tender document, the personnel involved in the estimating and adjudication processes will need additional information which includes the following:

- Manpower requirements
  - Own operatives – characteristics, mood and productivity of directly employed operatives and a sound average value of output.
  - Sub-contractors – performance, capabilities and conditions of working.

  - Present workload – availability of key management personnel to undertake the project successfully.

- Materials – cost and availability on demand.
- Plant – availability, effectiveness and cost of hire/purchase.
- Finance – ability to borrow, cost of borrowing and client's ability to make prompt payment(s).
- Familiarity with project – affording competitive advantage over rivals.
- Site conditions/atmosphere – dependent on weather conditions and the efficiency of management in site planning.
- Market conditions – influence level of competition and profit margins.
- Project overhead – dependent on the number and calibre of management and administrative staff and equipment allocated to the project.
- Rivals move – the likely opposition and how they do react to given situations.
- Risk factors – best markup provided to cover any high risk and uncertainty identified.

The phases of the establishment of production costs normally revolve around:

- The receipt of tender documents and the initial examination and decision whether or not to tender.
- On affirmation of the decision to tender, initial meeting to establish the programme and responsibility of key players.
- Periodic reviews of the tender (under preparation) to confirm progress and sufficiency of input.
- The compilation of the estimated cost.
- The adjudication which converts the estimated cost into a tender price.

The tender documents which contractors may be asked to price include bills of quantities, drawings and specification and schedule of rates. These tender documents should adequately describe the scope, quantity, quality and position of the work. Apart from reflecting the scope of the works as accurately as possible, the prime objective of the tender documentation must be to provide a basis for the submission of unqualified bids. To do this the tender documentation must provide a precise set of instructions to tenderers. Particular items which need to be identified in such instructions include:

- The mode of presentation of rates and prices in the bills of quantities.
- A list of the information that must be submitted by contractors with their tenders.
- Alternative construction programme and tender price relating to it.
- The method of dealing with errors and misunderstandings of the client's requirements by tenderers.

The client will get best value for money when full, clear and accurate information is produced to enable bidding contractors to tender accurately for the construction project. Poor documentation and information could mean that bidding contractors will not clearly identify the requirements of the project. In such a situation, the client may not achieve the best value for money and hence, he or she and the contractors will be taking unreasonable risks working to inadequate tender documentation. (See a comprehensive discussion of estimating and tendering techniques in the author's *Understanding tendering and estimating*.)

Generally contracting is a risky business and the more these risks are allowed for during the establishment of production costs the better. For this reason, contractors take all known risks into account in their tender submissions. Nevertheless, very often a contractor will have to consider whether his or her tender can stand competitively with other contractors if all contingent risks are allowed for in a higher bid. The competitive elements in tendering sometimes dissuade contractors not to price all known project risks adequately. Therefore, in order to give a high standard of service at a reasonable price and optimum profit, contractors plan construction projects adequately not only in the production process but also in terms of financial control. The form this planning takes in practical terms are (a) the budgetary control (discussed in Chapter 10) which facilitates periodic comparison between estimated and actual performance and (b) the institution of corrective measures when results are unsatisfactory.

8.7 TENDER APPRAISAL

On receipt of competitive tenders, they are evaluated and reported to the client with a recommendation for consideration. The tender evaluation is conducted by the quantity surveyor who checks that the following are in order:

- All pages in the bills of quantities are intact and that all items have been priced in the prescribed manner.
- A short line drawn through the cash column of items the tenderer did not intend to price.
- Bills of quantities contain neither unauthorised alterations nor qualifications.
- Any corrections notified during the tender period have been made to the relevant parts of the bills of quantities.
- The bill extensions and collections are arithmetically correct and have been transferred to both summary page and form of tender.
- The figure in the summary page of the bills of quantities agrees with the tender figure.
- The pricing is satisfactory, uniform throughout and the unit rates could be used as a fair basis for the valuation of variations.
- The prices for the same items of work in different sections of the bills of quantities are consistent throughout.
- Reasonable percentage and lump sum additions for profit and attendance on prime cost (PC) and provisional sums for specialist sub-contractors and suppliers have been made.
- Percentage additions made to the prime cost of daywork items (labour, plant and materials) are acceptable.

Example 8.1 Tender report

UPPER DAGNALL STREET DEVELOPMENT

REPORT ON TENDERS

1.0 INTRODUCTION

This is a report on tenders received for the above development project and a recommendation for the award of contract.

2.0 INVITATIONS

2.1 Invitations to tender were issued to six contractors on (date)

2.2 Invitation to tender stated that:
- Tenderers should submit a fixed price tender
- Tenders should be submitted to the client's offices by noon Friday (date)
- Tenders should be submitted for 78 weeks production period.

3.0 TENDER RESULTS

3.1 All six tenders were returned within the stipulated period and opened at the offices of the client on Monday (date). Results of tenders received are summarised as follows:

Name of tenderer	*Tender sum*
1. Fast Builders Limited	£2,728,043.14
2. Azec Construction Limited	£2,779,065.93
3. Top Construction Limited	£2,804,962.57
4. Church Construction Limited	£2,875,423.30
5. Supreme Builders Limited	£2,931,173.62
6. Harvey and Sons Limited	£2,986,814.08

3.2 From the above analysis, it can be seen that the tenders are competitive. The lowest tender is only 1.87% below the second lowest tender and it is also within our last cost report of £2,750,000.00 dated (date).

3.3 Given that the second lowest tender is approximately £51,000.00 higher than the lowest, it is considered not appropriate at this stage, to examine any other tenders but that of Messrs Fast Builders Limited.

4.0 EXAMINATION OF MESSRS FAST BUILDERS LIMITED'S TENDER

4.1 The priced bills of quantities submitted by Messrs Fast Builders Limited have been examined arithmetically and technically and we report as follows:

REPORT ON TENDERS (*Cont'd*)

4.1.1 Messrs Fast Builders Limited's tender contains an arithmetic error of £1,200.00 within the general summary.

4.1.2 The error if corrected, would increase Messrs Fast Builders Limited's tender by £1,200.00 to £2,729,243.14.

4.1.3 The error was brought to the attention of Messrs Fast Builders Limited who confirmed that they stood by their tender.

4.1.4 The general level of pricing is, in our opinion, fair and reasonable; it is consistent throughout with no serious anomalies.

4.1.5 Messrs Fast Builders Limited's tender contained no qualifications within the priced bills of quantities or within their covering letter submitted with their tender.

5.0 CONCLUSION AND RECOMMENDATIONS

5.1 It is clear that both tenders submitted by Messrs Fast Builders Limited and Messrs Azec Construction Limited are competitive with *'difference in price of less than 2% between the two companies'*. It also compares well with our comparable estimate prepared on (date).

5.2 Accordingly, Messrs Fast Builders Limited's tender in the sum of £2,728,043.14 is recommended for acceptance by the client.

- The preliminaries section of the bills of quantities has been priced individually and fixed and time-related charges are clearly shown.
- The tender contains no serious mistakes that (on notification) will force the tenderer to withdraw the tender.
- The basic rates (in cases of contract subject to price adjustment) of materials are reasonable.

The above technical check is aimed to ensure that all anomalies are clarified and agreed before the client is committed to contract. Failure to do so creates problems and confusion once production has started on site. As part of the evaluation procedure, it is at times, inevitable that a meeting must be held with one or more of the tenderers to clarify any points discovered in the appraisal. It is also important that any agreement reached on any matters affecting the eventual contract is separately recorded in writing for incorporation into the contract documents. The tender appraisal is completed by submitting a report to the client setting out all the actions taken to scrutinise, check and correct the tenders and giving conclusions and recommendations regarding which tender to accept.

8.8 TENDER REPORT

A tender report contains information on tender evaluation, general technical comments and recommendations. The report is forwarded to the client with a recommendation to assist him or her in decision making (see Example 8.1). The form and contents of the tender report varies from project to project, dependent upon the size and complexity of the project, but usually includes the following points:

- Tender sums for which each tenderer is prepared to execute the construction project and tender period when this has been made an object of competition.
- Arithmetical or abnormalities and/or inconsistencies found in the lowest (and at times the second lowest) tenderer's bids.
- The quantity surveyor's technical observation on pricing level, method or policy and quality of pricing.
- Details of any qualifications to the tender and their financial and contractual effect on the project.
- Conclusions and recommendations for acceptance or rejection.

When the client receives the tender report, he or she must be in a position to accept the recommended tender within the period for which the tender is open for acceptance or otherwise. On acceptance of the quantity surveyor's recommendations, and prior to giving the building contractor possession of the site, a contract should be drawn up and signed. In addition, the design team members should be instructed to proceed to the subsequent stages in the construction process.

8.9 CLIENT'S DECISION

On receipt of the tender report, the client may accept the design team's recommendations and give an order that production should start. Conversely, he or she may instruct that the tender sum should either be reduced (where the tender figure is too high and a reduction needed) or increased (where the tender figure is too low or an additional finance has been arranged for additional works). Where the client orders a reduction or addition to the tender figure, the design team achieves this by preparation of the following documentation for pricing.

8.9.1 Bills of reduction

Bills of reduction are prepared when the lowest tenderer's figure is too high and/or above the client's cost limit and the client requires a reduction of the tender figure. This reduction is obtained by altering the measured work contained in the bills of quantities or contract documentation and the alteration often involves complete omissions, changes in specification levels and/or quantities. In the process, the size/number of an expensive material/component may be reduced or omitted and a cheaper substitute added in replacement. For example, cheaper facing bricks, internal finishing items, plumbing items, ironmongery and so on may be substituted for expensive types contained in the bills of quantities. The presentation of the bills of reduction should be in a format understood by the building contractor and it is therefore necessary that the details of the varied items of work (i.e. what is being omitted from the bills of quantities and what is to be added back) are clearly shown (see Example 8.2 for a sample of the bills of reduction).

8.9.2 Addenda bills

Addenda bills contain items of additional work which are required to the original design and are added to the contract after completion of the main bills of quantities.

8.9.3 Correction of errors

Generally, a tender sum is a legal offer made by a contractor to carry out work and, hence, is not subject to an adjustment. This figure will only change by reason of variations to the contract. Therefore, errors in tenderers' bids are dealt with in two ways in accordance with section 6 of the Code of Procedure for Single Stage Selective Tendering. The first approach states that when errors are found in a tender, the tenderer is informed of the errors and then asked to confirm or withdraw the tender. The second course of action is to inform the tenderer of the errors and enquire if the tender is to be confirmed and if the genuine errors are to be corrected. If the tenderer elects to correct the genuine errors and this action makes the tender (under consideration) higher than the second lowest tender sum, that tender is set aside and, instead, the second lowest tender is considered. Example 8.3 shows how arithmetical errors are corrected in a priced bills of quantities where, notwithstanding the errors, the tenderer has elected to stand by the tender.

8.9.4 Notification of tender results

After tenders have been received and evaluated, the client should be in position to either accept a tender, reject all tenders or accept a tender subject to agreed modifications. Whichever the case, all tenderers should be notified in writing and supplied with a list of the tender figures. However, the successful tenderer's actual figure is not disclosed, but as each tenderer knows his or her own tender figure, an examination of the figures will enable each tenderer to tell how he or she stood in relation to the rest. An example of a tender notification for unsuccessful tenders for a hypothetical project is given in Example 8.4.

Example 8.2 Bill of reductions

					Omissions	Additions
	UPPER DAGNALL STREET DEVELOPMENT					
	BILLS OF REDUCTION					
					£	£
	ITEM 1					
	Substitution of coronet light Red stock facing bricks for hand made Sandringham Red.					
	OMIT					
	Facing brickwork; hand made Sandringham Red; mortar mix 1:1:6; stretcher bond; finished with weathered joint; in Clause F110B					
A	Walls; half brick thick; facework one side (as bq item 6/2/F).				5,794.97	-
B	Closing cavities; 75 mm wide; brickwork; half brick thick, vertical (as bq item 6/2G).				405.21	-
C	Ditto, brickwork 175 mm wide; horizontal; entirely of headers laid flat (as bq item 6/2/H).				89.74	-
	ADD					
	Facing brickwork; coronet light Red stock; mortar mix 1:3 SR; stretcher bond; finished with flush joint; as Clause F110A					
D	Walls; half brick thick; facework one side.	149	m^2	27.00	-	4,023.00
E	Closing cavities; 75 mm wide; brickwork; half brick thick; vertical.	67	m	4.75	-	318.25
F	Ditto; brickwork 175 mm wide horizontal; entirely of headers laid flat.	14	m	3.65	-	151.10
					6,299.92	4,392.35
	Net omissions					
	Carried to Summary			£	1,897.57	

Example 8.3 Correction of an error in a price bill of quantities

		£	
(a)	Tender sum:		
	Builder's work	1,457,255.14	
	Preliminaries	295,788.00	
	PC and provisional sums	975,000.00	
		£2,728,043.14	

However corrected errors makes this £2,729,243.14

(b)	Revised value of builder's work:		
	Corrected tender		2,729,243.14
	Less		
	Preliminaries	295,788.00	
	PC and provisional sums	975,000.00	
			1,270,788.00
	Value of builder's work		£1,458,455.14

The corrected builder's work is more than it should be by £1,200.00 therefore, the following error adjustment is required:

(c) Error adjustment: £

$$\frac{\text{Error}}{\text{Corrected value of builder's work}} \quad \frac{1,200.00}{1,458,455.14} \times 100 = 0.0823\%$$

Therefore to arrive at the tender sum, all unit rates in that builder's work are subject to 0.0823% reduction. This may further be demonstrated as follows:

	£
Corrected value of builder's work	1,458,455.14
Deduct 0.0823% =	1,200.31
	1,457,254.83
Add sum of preliminaries, pc and provisional sums	1,270,788.00
	£2,728,042.83

What the above calculations mean is that during the production phase, all builders work included in the interim valuations and any variation valued at the bill rates are subject to a reduction adjustment of 0.0823%.

Note: The difference of £0.31 in the above example is the result of limiting the percentage reduction to four decimal places.

Example 8.4 Tender notification

Upmarket Designers
177 Baker Street
London
SW12 7BE

UPPER DAGNALL STREET DEVELOPMENT

NOTIFICATION OF TENDER RESULTS

Date:

Dear Sirs

Thank you for your tender dated (date) in respect of the above project. We regret to inform you that your tender was unsuccessful on this occasion. However, for your information, a list of the tenders received is as follows:

(a) £2,728,043.14
(b) £2,779.065.93
(c) £2,804,962.57
(d) £2,875,423.30
(e) £2,931,173.62
(f) £2,986,814.08

Yours faithfully

Partner

SUMMARY

In this chapter we have discussed the contractor selection procedure and the establishment of the contractor's price for the project's production. The governing factor in these twin processes is time, as effective timing of certain construction decisions can lead to financial savings for the client. There are many methods of contractor selection, however the method adopted should reflect the client's development strategy (i.e. the procurement of a good-quality construction product economically, safely and expediently).

The building contractors who consent to bid for a construction project production do so under several motives. While profit may be the prime motive, some contractors bid in order to secure further business when their order books are low and/or to achieve the planned level of turnover. As the bidding process consumes time and money, contractors put a lot of effort into the bidding process in order to win more contracts and thereby achieve the expected turnover and return to enable them to finance the bidding cost.

FURTHER READING

The Aqua Group 1989 *Pre-contract practice for the building team.* Blackwell Scientific Publications.

Franks, J. 1991 *Building contract administration and practice.* Batsford-CIOB.

Kwakye, A.A. 1994 *Understanding tendering and estimating.* Gower.

Rougvie, A. 1988 *Project evaluation and development.* Mitchell-CIOB.

Turner, D.F. 1983 *Quantity surveying practice and administration.* George Godwin.

Willis, C.J.; Ashworth, A. 1990 *Practice and procedure for the quantity surveyor.* BSP Professional Books.

CHAPTER 9

PRODUCTION ADMINISTRATION

9.1 INTRODUCTION

Action on receipt, evaluation and acceptance of a contractor's tender is concerned with the preparation for the commencement of production. This preparatory activity is undertaken by both the contractor and the client's professional advisers and takes diverse forms dependent almost always upon the size and complexity of the construction project, availability of information and preparatory activity carried out at the pre-tender stage.

The preparatory activity is essential, as although the client's professional advisers may have reached this stage of the construction project, they must ensure that appropriate measures are adopted before the contractor is appointed. The rationale behind this move is that, after the contractor's appointment, any unfinished preparatory work which delays the commencement of production or disrupts the contractor's programme of work is likely to expose the client to additional costs and a possible construction dispute. The preparatory work, therefore, ensures that when the contractor has been appointed, production on site should then progress without any major interruptions.

The contractor, on the other hand, needs time to mobilise his or her resources. Like the client's professional advisers, the contractor's preparatory action prior to site production is to ensure that all the essential resources and notices required for the commencement of site operations are in place. The above is a clear indication that the preparatory work undertaken by the project team are essential to avoid delays and, hence, drive the project to a successful completion. For this reason, any amount of time spent in this regard is not time wasted as the benefits of this action can be enormous.

9.2 CLIENT'S PROFESSIONAL ADVISERS' PREPARATIONS

At this stage, before the commencement of production on site, it may be assumed that the competition or negotiation has run its course and come to an end and that decision on the quantity surveyor's tender report has been made.

Therefore, it may be further assumed that the client has accepted an offer (in most cases the lowest tenderer's bid) and has instructed that a contract should be placed with the successful tenderer. However, prior to signing the contract, the client's professional advisers usually have a second check to see if the following are in place:

1. *A clear site/land:* Where the site/land is occupied, a check is made to determine whether those occupying it have been successfully decanted or relocated. Where the site/land has yet to be cleared at this stage, then it needs a serious consideration as decanting occupants of a site/land can be time consuming and can delay the commencement of construction projects. On occasions, this process requires a court action to enforce the move. Also, in situations where the site/land has not been cleared before contract and commencement of the works, the contractor's progress had been frustrated and clients have ended up paying the contractor's claim for financial reimbursement resulting from uncleared site.
2. *Necessary planning approval:* Giving the contractor possession of site without the necessary planning approvals can be both frustrating and costly. Valuable construction time may be lost through delays in obtaining the necessary approvals, licences and permits.
3. *Availability of funds:* The client must be able to confirm at this stage whether he or she has arranged the necessary funds for the construction project and/or whether his or her financial position has not altered since the inception of the project.
4. *Insurances:* A check is carried out to ensure that the necessary insurance cover can be procured for the project, especially where the element of risk is high.
5. *Statutory authorities:* Tardy response by statutory authorities can be a cause of delays and hence, a check is made to ascertain whether they have all the information required on the proposed construction project and are ready to carry out their part of the works (e.g. diversions of existing services, provision of crossover and the like) when required.
6. *Nominated/named sub-contractors:* The architect checks with the successful contractor to ensure that he or she has no valid objection to working with any nominated/named sub-contractor or supplier mentioned in the contract documentation.
7. *Contract documentation:* All necessary construction information such as copies of drawings, specification and bills of quantities are issued and/or made ready for handing over to the contractor.

Analysis of tender

The client's professional advisers carry out analysis of the tender as required by the client and, at the same time, update and finalise the cost plan and establish a proper basis for post-contract cost monitoring and reporting.

9.3 CONTRACTOR'S PREPARATIONS

As soon as the client's tender acceptance letter (or, at times, letter of intent) is received and prior to signing the contract documents/agreement (which at times occurs several weeks later), the successful contractor starts the planning and mobilisation action. The planning required to be undertaken at this stage covers site layout, site operation and the organisation of site staff. These should proceed concurrently, as decisions made on one may affect the others.

9.3.1 Planning generally

Planning is a management process which enables successful running of many business activities, including construction projects. In construction projects, planning is concerned with the various methods of arranging, procuring and employing money, materials, men and machines (all of which are known as resources) to carry out the day-to-day construction operations on site. It is also a practical attempt to foresee problems before they arise and, hence, avoid events which might prevent the attainment of stated project objectives, namely the completion of construction project within a fixed timescale, safely, economically and to specified quality levels.

9.3.2 Construction project planning

Construction project planning is primarily about thinking ahead and therefore entails predicting the working method to be employed, resources to be utilised, when certain events should happen, duration of activity and so forth. There is always a certain amount of risk associated with all predictions and one approach to the reduction of the risk is to get as many people as possible involved in the planning. For this reason, the contractor creates a broad planning dialogue in which persons such as managers and senior personnel (especially those who will be closely associated with the project) are drawn into the planning process and are also made to feel responsible for their respective contributions. In addition, as an effective line of communication is also of great importance at this stage, the contractor arranges a formal pre-contract meeting to discuss the various contract requirements, production planning, delegation of responsibilities, establishment of lines of communication and so on. Generally, at this stage, production planning is concerned with completing the construction project in the shortest possible time compatible with quality, technology, economy, legal, social and safety parameters.

Planning must also build on a clear division of the project into stages. For this reason, the contractor commences the planning process by analysing the project and defining the tasks to be executed. What determines the size of individual tasks are factors such as manageable proportions, resource availability, speed of production, room for manoeuvring, external circumstances and the like. In the process, the contractor transforms the drawings and measured works section of the bills of quantities into a number of construction operations, tasks or activities. These operations, tasks or activities form the basis of a method statement and are also incorporated into the contract programme.

9.3.3 Method statement

A method statement is a detailed assessment and narrative of the construction methods to be adopted on a construction project. It involves turning the drawings and bills of quantities into a number of construction operations, tasks or activities and detailing a suitable method of carrying out every operation, the duration and the combinations of equipment and labour power. This should correspond in all respects to the contract programme.

In this technological age, computer software is available for use to save time and resources required in carrying out this exercise. A contractor who has this facility therefore, normally feeds information on the construction project into a computer and it expediently works out programmes, critical paths, float and so forth.

Generally, the information contained in method statements is used for compiling the bid estimate and, hence, is normally prepared at the tender stage (see Table 9.1). For this reason, it would be appropriate to assume that the preparation of a pre-contract method statement entails adjusting, updating and finalising the pre-tender method statement for use in planning and programming the works. However, where no method statement exists, one is prepared by the contractor on the award of the contract, and decisions on labour and plant requirements are made by those who will be responsible for managing the project and providing resources (e.g. planning/contract manager, programming personnel, site manager and plant manager). The method statement should contain all the relevant information on the operation such as work stage, quantity of work, method of construction, output,

Table 9.1 Method statement of Upper Dagnall Street Development

Description of items	Quantity	Details of method	Plant	Output per week	Plant labour involved	Period required
Excavate pipe trench	500 m	Excavate backfill plant	Backacter	500 m	4 labourers	1 week
PVC pipes	500 m	Lower by hand	Nil	250 m	6 labourers	2 weeks
Basement excavation	4,000 m^3	Excavate direct load to lorry	Backacter and lorry	2,000 m^3	2 labourers	2 weeks
Basement reinforcement	5,000 kg	Supplied cut and bent	Nil	1,666 kg	2 steel fixers	3 weeks
Basement concrete	400 m^3	Site mixed	14/10 mixer	100 m^3	6 concretors	4 weeks

plant summary and duration. It must also be in a format which depicts the general sequence in which the various tasks are to be carried out, and if it is to be presented to the client, it will also include safety aspects of production.

On completion, the method statement should be able to convey to its users the planned execution of the works. Furthermore, it should prove an indispensable tool for use by the planning and programming personnel in the preparation of the contract programme and also by the site manager for organising and managing the daily site operations.

Besides pulling together managerial experiences on production methods and effective utilisation of resources at tender stage, the method statement at the pre-contract planning stage offers the contractor other benefits, as follows:

- It coordinates the efforts of the contractor's site management team in their determination to get the project completed efficiently, economically and on time using an agreed method or sequence of work.
- It provides each section of the contractor's establishment with detailed information on construction.
- It enables data concerning new construction technique and effective methods of handling materials/components to be included in the project planning.
- It allows a realistic comparison of output and duration of different methods, machines and combination of personnel/machines and, hence, facilitates the choice of optimum production method.
- It permits the plant requirements for the construction project to be scheduled for procurement by the plant department.
- It acts as a feedback system for the contractor's estimating department.

9.3.4 Contract programme

As soon as the contract is awarded, the successful contractor prepares a suitable programme to enable the works to be carried out in an orderly and efficient manner. The contractor's programme normally consists of a statement of intended construction operations set in a logical order to assist the smooth running of the project. The programme also facilitates the identification and avoidance of potential production problems. A careful study and planning of the various construction operations are prerequisite for the production of a suitable programme which shows all the operations (e.g. what activity should happen, when it should happen and by whom it is carried out; how long it takes and how all activities dovetail together – see Figure 9.1). In effect, the programme must show sufficient detail in terms of timing and duration of operations, dates for delivery of materials/components, labour resource requirements and sub-contractors site commencement and completion dates. It should also identify the critical path.

Generally, the construction programme is developed by the contractor from the programme prepared at the pre-tender stage and serves the following purposes to the contractor's site management team:

Fig. 9.1 Master programme

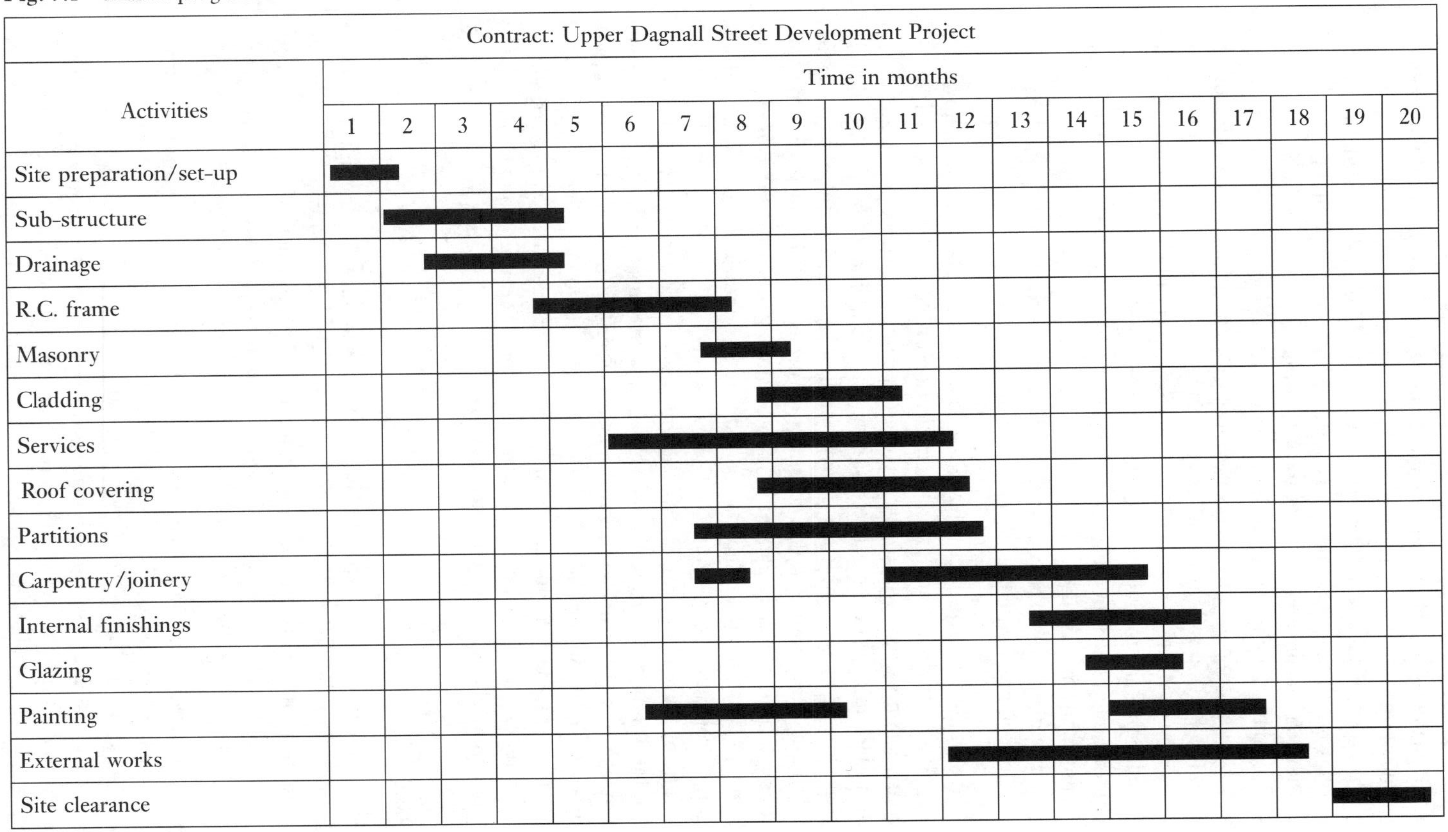

Contract: Upper Dagnall Street Development Project																				
Activities	Time in months																			
	1	2	3	4	5	6	7	8	9	10	11	12	13	14	15	16	17	18	19	20
Site preparation/set-up																				
Sub-structure																				
Drainage																				
R.C. frame																				
Masonry																				
Cladding																				
Services																				
Roof covering																				
Partitions																				
Carpentry/joinery																				
Internal finishings																				
Glazing																				
Painting																				
External works																				
Site clearance																				

- Depiction of visual instruction and forming the basis of controlling all site operations.
- Provision of yardstick for progressing, controlling, reviewing and costing.
- Showing the sequence necessary for carrying out an operation, the duration and the total output required of resources employed.
- Providing the client with an indication of its periodic financial commitments.
- Enabling the non-productive time of both men and machines to be identified, controlled and minimised where possible.
- Exposing, where foreseeable, likely difficulties and delays in the future and facilitating the institution of corrective measures to overcome them.
- Discouraging design changes as the programme depicts the natural consequences flowing from actions.
- Indicating at what stage, during production, the contractor would require design information from the architect.

The overall/master project programme is a document which breaks down site activities and hence, enables the identification of the total work content of the construction project. Moreover, it covers the entire contract period and includes all operations in broad terms. The resulting set of operations are quantified and then phased using either simple bar charts (for simple construction projects) or network techniques (for more complex construction projects). Where bar chart technique is adopted, time is usually plotted in months and weeks; dates and contract week numbers are entered and holiday periods are also indicated to show lost production in the programme.

The overall/master project programming is carried out by the contractor's central planning department in consultation with the site manager and contracts manager, but the detailing is sometimes prepared by the site manager under the supervision of the contracts manager. The consultation and liaison with both the site manager and the contracts manager is important as they will be responsible for interpreting the programme into a series of short-term programmes during the production phase of the construction project

9.3.5 Site layout

Site layout involves the study, planning and organisation of the unused areas of site around the proposed development to accommodate the contractor's construction equipment, materials and buildings for use in the execution of the construction works. The siting and/or arrangement of temporary buildings, plant and so forth may be influenced by one or more of the following factors:

- Type of construction project (e.g. low-rise or high-rise structure) and method of production envisaged.
- Size, shape and extent of proposed works.
- Size and shape of site and the area to be covered by the proposed works.
- Position of existing obstructions (e.g. existing services and structures) and hazards such as gas mains and underground or overhead electric cables.

- Accessibility to and restrictions within the site.
- Type and size of construction plant and equipment planned for use.
- Proximity of site boundary to existing buildings.

A site layout may be worked out using a plan of the site showing an outline of the proposed building(s), roads, paths, pavings and the like. The proposed route of main service runs is usually marked to avoid placing temporary building or plant over this route. Existing buildings, services, trees and other obstructions on site should be noted on the site plan. In particular, the names of all statutory authorities whose services run across the site should be noted, and, where necessary, they should be contacted for details and location of any services that cross the site but have not been shown on the site plan.

One of the primary considerations for site layout is the need to keep the construction production moving at all times by maintaining a means of access to the site and facilitating an adequate flow of traffic. However, of most importance is the consideration given to the most critical or difficult operation on the site, and this should be the prime factor deciding the layout. In addition, there is also the need to pay regard to methods of unloading during the operation and the vehicle turning circle. The following group of items will need to be considered in the development of site layout.

Temporary buildings

The number, type and size of temporary buildings depend on the extent of the construction project, the client's requirements, minimum statutory requirements and contract period. Temporary buildings are now available in many forms, sizes and makes; however, the contractor's selection should aim at provision of a proper working environment, and privacy to reflect the status of the occupant.

Temporary buildings can be hired or purchased and contractors normally consider factors such as size, ease of erection and dismantling, adaptability, weather tightness, life expectancy, thermal conductivity and ease/cost of transporting to and from site. When taking decisions regarding their acquisition the temporary buildings which are required on most sites include the following:

1. *Offices:* The siting/arrangement of site offices takes account of the interrelationship between members of the site management team. In the UK, the statutory requirement for each office worker is 3.7 m^2 of floor space and the size of office space provided will depend on the number of site staff. The office should have separate sanitary conveniences for each sex at the rate of one per twenty-five persons and washing facilities should also be provided. In addition, offices should be equipped with desks, chairs, plan chests, storage containers, telephones and the like where necessary. Provision should also be made for adequate fire escape routes, first aid, lighting and means of heating.

Generally, offices are provided to the following groups of site staff:

(a) The site manager/agent. This office provides all site/construction information and, hence, is the control centre of the site. Therefore, ideally, it must be placed at a vantage point so as to provide a good view of the site and also to facilitate checking, supervision and security.
(b) General staff – these include site engineers, surveyors, administration support clerks, safety officers and so on.
(c) Clerk of works, resident engineer and/or architect (where stipulated) – a separate telephone service may be required, together with adequate mess facilities.

2. *Welfare building:* This is the building to which operatives and site staff retire to eat their meals during the lunch break. There must be facilities for warming meals and boiling water and, for this reason, on large sites, canteen facilities are sometimes provided. The size and extent of the site canteen and mess facilities for operatives is mainly dependent upon the extent, location and number of operatives employed on the project. The siting of the canteen is normally influenced by the time taken by operatives to get to the canteen from their places of work, position of site services (gas, electricity and water), access for delivery of food stuff and removal of kitchen waste.
3. *Drying rooms:* Drying rooms and lockers are provided for drying and storing operatives' clothes. Ideally these rooms should be located close to the canteen so that clothes can be dried out during lunch break where necessary. Moreover, these rooms should be adequate to enable operatives to put on and remove protective clothing.
4. *Sanitary conveniences:* Sanitary conveniences and washing facilities are also provided for the site operatives. The number of sanitary conveniences provided depends upon the maximum number of operatives on site at any one time, and again its provision is at the rate of one per twenty-five operatives. The siting of the sanitary conveniences is to some extent dependent upon the position of available drain connections and the facility normally includes hot and cold water, towels and soap. On sites without drainage connection, portable chemical toilets (cleaned out weekly) can be provided for usage.

Material storage areas

Material storage and work areas for site operations will need to be marked out clearly for the various materials and operations. Storage or a secure store is necessary for valuable items which may be stolen if not adequately stored away and protected. Furthermore, a weather-proof store will be needed for materials which may deteriorate due to the effects of weather. Open storage areas are also required for materials such as bricks, timber, scaffold, precast concrete units and drainage goods.

Location of plant

The location of major items of static plant such as hoists, tower cranes or the space for parking mobile plant are given careful consideration. The storage compound for tower cranes, for example, are so arranged that its radius covers the maximum area of the site. This arrangement enables it to be used for unloading items from delivery lorries into storage areas as well as hoisting items from store onto workplaces or the structure. Particular regard is also paid to erection and dismantling of the tower crane. In addition, rights of airspace and over-rail agreements are considered in relation to site boundaries where necessary.

The other major item of static plant is the concrete/mortar mixing equipment. These are always an attraction to vandals and therefore night storage areas and/or watching will be required for these items of plant.

Temporary roads, hardstanding and access

The provision of temporary roads for the movement of plant and materials is an important issue which requires care and attention. Ideally, where possible, the temporary roads are planned to follow the layout of the permanent road to effect savings on resources in the road construction. Hardstanding may be required for static plant such as tower cranes undertaking lifting operations. Moreover, the provision of sufficient car parking should not be overlooked by the site planner as most operatives, management, site staff and visitors normally arrive on site by car and, hence, require parking space on site for their vehicles.

Sundry points

The following items also require consideration:

- *Stand pipes* should be located close to a mixer or where most required.
- *Site name boards* should be located in prominent positions to identify the site. However, often the positions of site name boards are dictated by the client.
- *Skip location* should be placed close to the works but should be accessible to facilitate exchange at regular intervals.
- *Vehicle wash areas* are required to prevent the site mud spreading onto the highway.

9.3.6 Schedule of resources

Prior to commencement of site operations – and in addition to the method statement, contract programme and site layout plan – the contractor prepares a resources schedule. The schedule is composed of plant, labour and material requirements and the following are considered in most schedules of resources.

Plant schedule

The plant schedule consists of a list of major plant items required for the construction project. At this stage of the construction process, it is essential to reconcile the cycle of equipment usage with decisions made at the pre-tender stage (i.e. the method statement). Decision is also made on whether to hire or purchase an item of plant/equipment for the project (especially where the contractor does not own the required plant/equipment).

Staff schedule

The staff schedule comprises a number of key management, technical and clerical personnel allocated to a construction project. This schedule is normally presented in a form of site organisation structure chart to keep all concerned informed of their authority, responsibility and the channels of communication. Where, owing to under-capacity, suitable staff cannot be found within the contractor's establishment for the required positions, a decision may be taken to engage new staff on a temporary or permanent basis.

Labour requirement schedule

The labour schedule consists of a list of operatives required for the construction works, including gang sizes required for each of the site activities. This schedule could list the contractor's own operatives transferred from other sites or, on the other hand, operatives could be recruited from job centres or hired from specialist labour-only companies or sub-contractors.

Materials schedule

The materials schedule includes a list of all the main materials/components required for the construction project. The items listed have been extracted from the bills of quantities, drawings and the specification and are ready for orders to be placed. The various dates for placing orders may be obtained from a master programme and the contractor's buying department normally places orders as and when required with the suppliers who provided the most competitive quotation and/or whose prices were used in the tender estimate. The ability to supply the material/components on time and to meet the construction programme must also be a consideration.

Schedule of sub-contractor's work

This is composed of a list of nominated/named sub-contractor's and domestic sub-contractor's work. These are arranged in order of dates to show when a particular sub-contractor's input is required on site.

9.3.7 Sundry arrangements

As careful planning at this stage is essential for the smooth commencement and administration of the construction project, the contractor pays particular regard to the following:

1. *Detail drawing requirements:* All required drawings and any other construction information are scheduled, giving dates by which they are required. The contractor forwards this schedule to the architect immediately for action.
2. *Temporary site services:* Information is obtained on positions of existing underground services (e.g. gas, water, electricity) in order to protect them during the site operations. The contractor also arranges for the supply of all temporary services (e.g. water, electricity and telephone) required for the execution of the works.
3. *Insurances:* Arrangements are made for the procurement of all insurances required under the contract (e.g. employer's liability insurance, insurance against injury, loss or damage and so on).
4. *Licences, permits and notices:* Licences required for hoardings, gantry and the like and also permits required for road closures, skips and catering facilities on site are obtained. The contractor also gives the required notices to the area local authority and the police before the commencement of the works.

Generally, the objective of all the foregoing planning and preparatory work by the contractor is to provide the best working environment and conditions which leads to safe, efficient and economic production of the client's project. In many instances, the above moves are legal/statutory requirements.

9.4 THE FLOW OF PROJECT INFORMATION

In order to ensure that the project team is working in unison, the architect disseminates various project information through the medium listed below.

9.4.1 Pre-contract briefing meeting

Generally, the coordination production information and dissemination are important to all the project interfaces. Therefore, as soon as practicable after placing or signing the contract, the architect (as the project team leader) takes on the role of supervising, investigating and coordinating all project information. In order to perform this role effectively, he or she arranges a meeting (known as the pre-contract briefing meeting) of all the principal members of the design team, together with the contractor's site management team. At this initial stage, the architect's objective is to build up a closely knit, smoothly operating group. As the site is not normally set up at this early stage, the first meeting is generally held in the offices of the architect. Traditionally, the architect chairs the pre-contract briefing meeting and those invited to attend would normally include:

Employer's representatives, i.e.

- architect, quantity surveyor, structural engineer, services engineer and clerk of works.

Contractor's representatives, i.e.

- contract's manager, site manager and project surveyor.

Where necessary,

- principal nominated sub-contractors and principal nominated suppliers.

All the above personnel, and any others invited, are sent details of the meeting such as the place, date and time, together with a proposed agenda which may include the following items for discussion:

1. *Introduction:* Under this item, it is intended that all present at the meeting will introduce themselves. In the process, each person will give his or her name, company represented and function.
2. *Meetings and minutes:* There will be a general discussion and agreement between the main parties on who is to be the chairperson of all future meetings. Also, the responsibility for recording, producing and distributing minutes, and to whom they will be sent, will be decided at this meeting.
3. *Contract particulars:* Under this item, the architect will confirm the client's acceptance of the contractor's tender, the tender figure, type of contract, contract period, commencement and completion dates, and will provide any other relevant information. If the contract has yet to be signed, the architect will indicate when it is expected to be signed. Furthermore, the position regarding various approvals and notices required will be discussed.
4. *Contractor's programme:* Where a programme for the work has already been produced, the contractor will distribute copies to all present. Thereafter, a general discussion on the programme will ensue; the contractor will take note of all necessary changes and a date will be agreed for the issue of the revised and finalised programme. However, where a programme has not been produced, the contractor will give an indication of the date on which it will be ready. In addition, the contractor will be asked to explain briefly the planned sequence of executing the main operations of the project, the methods, approximate dates for completion of main elements and any risks/difficulties envisaged.
5. *Site boundaries:* The architect will confirm the position of the site boundaries to the contractor. If the contractor has prepared a site layout showing positions of the temporary buildings, plant and so on, this will be discussed and any problems and constraints resolved. The type, size and position of signboards will be agreed and also any additional information required on the signboards will be clarified.
6. *Site inspections:* The architect (where not to be resident on site) will indicate and agree with the contractor days/times for a weekly site visit.
7. *Drawings issue:* Where the architect has any additional or revised copies of drawings and any other written construction information, these will be handed

to the contractor. Furthermore, the contractor may be informed of any drawings under amendment and what the amendments entail and when they will be ready for issue. The architect will also establish that the contractor has received sufficient drawings and other construction information to make a start on the project. Based on the programme, the contractor will request any outstanding further construction information such as drawings, details, specifications and schedules.

8. *Sub-contractor's/suppliers:* Under this item, the names of known nominated/named sub-contractors and suppliers will be passed to the contractor with an assurance that the list will be followed later by an architect's formal written instruction. Where specialist firms have yet to be nominated, or specialist work is out to tender, the contractor will be informed of the dates when decisions on these nominations will be made. Moreover, where the contractor intends to sublet sections of the work to sub-contractors (as they normally do), it will inform and submit a list of proposed sub-contractors to the architect for approval.
9. *Communications procedure:* Procedure for distribution of all correspondence and architect's instructions and number of copies required will be discussed and agreed. In addition, the status of oral instructions given on site by the architect, other consultants site instructions and clerk of work's instructions to the site manager will be clarified.
10. *Interim valuations:* Dates for the preparation of interim valuations for the whole contract period may be agreed. The architect will give the contractor an indication of the maximum period, from the date of the valuation, within which he or she is to expect payment from the client. Also the method of dealing with Value Added Tax (VAT), dayworks, fluctuations and valuation of preliminary items (where necessary) will be discussed and agreed.
11. *Insurances:* Under this item, the contractor will be reminded of his or her responsibilities under the contract with regard to obtaining insurance cover, and may be asked to submit for approval an insurance quotation with respect to employer's liability insurance. In addition, the contractor will be reminded of his or her obligation to provide evidence of premium receipts for the employer's liability insurance, and other contractual insurances, within a certain time period. Where the client is responsible for the insurance, as in the case of projects involving existing property, it is the architect's duty to check that this is in place and, hence, will press the client for assurances about this and production of evidence within a certain time period.
12. *Any other business:* Under this item, all present will be at liberty to introduce relevant matters for discussion, raise problems or provide additional information relevant to the project but not covered in any of the previous discussions.
13. *Site meetings:* Dates of the first and all future site meetings – place, time, length of notice required and who should be invited to attend – will be discussed and agreed. These site meetings are usually held monthly, but may be fortnightly or weekly depending on the rate at which construction information is required to be exchanged.

9.4.2 Types of site meeting

During the progress of the works different types of meetings are held in order to exchange and disseminate project information and to monitor and control the progress of the works. Every meeting thus performs a specific function and that function is directly or indirectly related to the successful completion of the project. These site meetings may be categorised as follows:

- Internal control or domestic meetings.
- Sub-contractor/supplier control meetings.
- Employee information meetings.
- Contract administration meetings.

The foregoing categories of site meetings are discussed fully under sections 9.6 and 9.7 and need not be dealt with here. However, after the pre-contract briefing, the project participants will continue to work on the information so far gained. Both the architect and the contractor will be working towards the provision of more project information, equipment or resources for commencement of production. The architect and other consultants will exchange and share production information and, if necessary, revise contract drawings. Moreover, the contractor, if he or she has not already done so, will deal with the required notices, permits, insurance and so forth and will mobilise the resources necessary for the commencement of the project.

9.5 SITE PRODUCTION

9.5.1 Setting up the site

At this stage, it is assumed that the contractor has taken possession of the site and has started the initial site preparation and setting out, followed by preparation of a site grid, recording of site levels, establishing the site boundaries and the like. After these initial operations, all the required temporary buildings, administration files and stationery, first aid facilities and so on will be brought onto the site. Also of great importance is the control of site communication process which involve the receipt, recording, distributing, storing and retrieving letters, drawings and other written information on site. This process also covers the determination of what telephone calls to record, the mode of recording and those not worthy of recording.

9.5.2 Project administration

The administration of the project is the collective responsibility of the project team, made up of the architect, quantity surveyor, structural engineer, services engineer and contractor (see Figure 9.2). It can be noted that, when comparing this figure with Figure 6.1 (p. 98), the names of five new personnel have appeared on the structure of the project administration. These personnel are listed below:

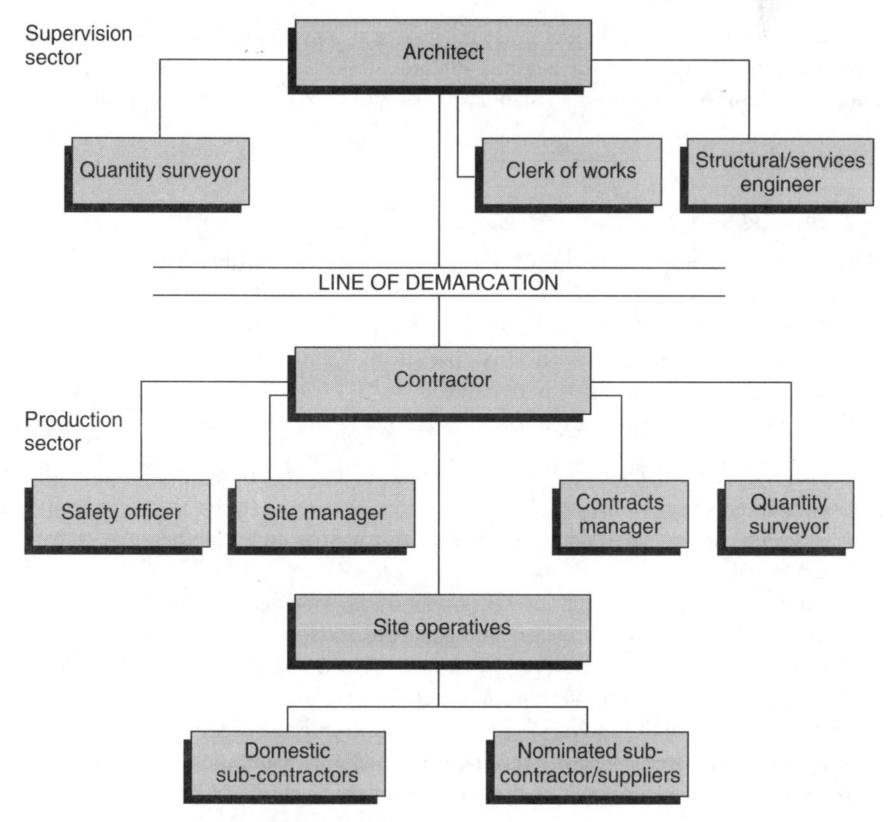

Fig. 9.2 Project organisation structure at the production phase

Clerk of works

The clerk of works is a quality inspector for a construction project during its production phase. He or she is usually appointed by the client, with the main duty of ensuring compliance with the contract in regard to the quality of materials/components and workmanship expended in the construction process. Ideally, the clerk of works should be appointed sufficiently early to enable him or her to become familiarised with the contract documentation and procedures before site production commences. Being on site most times, the clerk of works has the opportunity of inspecting any part of the works under construction any time before it is covered up. The clerk of works is responsible for checking the site grid and setting out, inspecting materials/components upon delivery and testing or submitting for testing any materials to be used in the works. In addition, he or she also informs the architect from time to time of any outstanding information required by the contractor as well as any discrepancy in the documentation. The clerk of works also endorses daywork sheets, assists the client's quantity surveyor with site measurement, and keeps records of progress of the project, working conditions and weather, delays, verbal instruction of the architect, site labour employed and visitors to the site.

Being in charge of quality matters, the clerk of works establishes the standard of workmanship laid down in the contract documentation and ensures that the contractor achieves the specified standard without any reduction in the quality of the work.

Contracts manager

The contracts manager is a contractor's representative responsible for the management of production of the construction project. The contracts manager is the key person upon whom the success or failure of the entire site production activities lies and, hence, takes an active part in the planning, programming and controlling of projects in his or her charge.

The contracts manager's responsibilities are many and may include the following:

- Preparing the site layout plan and organising the commencement of work by the programmed date.
- Examining the tender to determine the adequacy of working method and welfare facilities planned.
- Finalising appropriate order and safe methods of working; and allocating responsibilities to sub-contractors and trade supervisors.
- Ensuring that site production progresses as planned and instituting corrective measures if production targets are not met.
- Ensuring effective utilisation of contractor's time and resources on site.
- Preparing cost forecasts of site resources (e.g. labour, plant and materials), their utilisation and deviations for analysis by the quantity surveyor.
- Ensuring that all sub-contract trades are progressing smoothly.
- Issuing all necessary contract notices and letters.

Site agent/manager

The site manager is the contractor's employee and is in charge of directly managing the site operations. The site manager spends most of his or her time on the site talking to supervisors about technical construction matters, and is therefore mostly concerned with coordinating the people directly involved in the performance of construction tasks and welding together effective production teams. He or she is responsible for the day-to-day site operational decisions. These tasks are performed without reference to senior management at head office. Decisions are made on the basis of his or her own personal knowledge and often in response to some particular problem necessitating immediate action. In addition, the site manager is responsible for overseeing that all construction regulations are observed and reporting any defects in plant or equipment to the contracts manager. He or she also carries out periodic checks on the condition of site accommodation and plant, and chases up orders for material, plant and equipment. Above all, the site agent/manager is at the heart of the site operations.

Safety officer

The safety officer, a full-time employee of the contractor, is responsible for carrying out and regulating the contractor's safety policy. He or she is a trained and experienced person with detailed knowledge of all the appropriate construction safety legislations. The safety officer advises management on any legislative and legal requirements affecting safety in construction and any subsequent changes. He or she also advises on measures to be adopted to promote safe working methods and supervises the recording and analysis of information on injuries, damage and production loss. Above all, the safety officer must regularly assess and ensure the maintenance of the site safety standards.

Contractor's quantity surveyor

The contractor's quantity surveyor is the construction professional who carries out the quantity surveying functions in the contractor's establishment. The range of tasks performed by the contractor's quantity surveyor are somewhat different from those of the client's quantity surveyor. The contractor's quantity surveyor's roles vary from company to company in accordance with a particular company's size, policy and requirements. Generally, the contractor's quantity surveyor is more commercially minded and largely represents the contractor's interests in terms of the profitability of construction projects undertaken. For this reason, he or she ensures that appropriate measures are taken to make the construction project under his or her care a financial success.

In the management structure of some large contractor's establishment, quantity surveying is seen as a separate function/department under the direction of a senior and experienced quantity surveyor, who is normally a senior executive director. But the quantity surveyor's position is different in smaller contractor's establishments employing, for example, one or two quantity surveyors. Also, the size of the contractor's establishment influences the duties of the contractor's quantity surveyor. In smaller contractor establishments, the quantity surveyor is expected to possess a wider range of skills which enables him or her to deal with work which is normally outside the sphere of quantity surveying discipline (e.g. responding to queries such as why certain materials, plant, suppliers and sub-contactors are late in arriving on site or why a particular operation has taken longer than the time allocated in the tender estimates). But in large contractor establishments, the contractor's quantity surveyor undertakes a rather narrow or specialised range of tasks. However, whatever the role, the contractor's quantity surveyor must give an account of the financial success or failure of projects under his or her control. Depending on the requirements and policies of a particular contractor's establishment, the main responsibilities of the contractor's quantity surveyor include the following:

- Undertaking measurement, estimating and negotiations for new contracts.
- Preparing and agreeing measurements, interim valuations, final accounts and contractual claims with sub-contractors and the client's quantity surveyor.

- Collecting or recording cost information on site operations for use in the preparation of contractual claims, settlement of final account or for use in the preparation of future estimates/tenders.
- Identifying variations to the works and requesting the architect to issue formal instructions to cover the varied work.
- Providing assistance in the procurement and evaluation of sub-contractor's quotations and preparing sub-contract documentation.
- Preparing and comparing costs of alternative production methods of various site operations to enable the selection of the most economic method.
- Preparing periodic (usually monthly) financial reports (e.g. cost value reconciliation) and cash flow forecasts for management.
- Attending site meetings with client's representatives and sub-contractors and, where necessary, clarifying and ensuring that the contractor's contractual position is protected.

General

The above-mentioned personnel perform their respective roles on the construction project and contribute towards the achievement of the client's development aims. The architect, who is traditionally the project team leader, ensures that all the participants are pulling in one direction and any conflict which threatens the objective of the project is dealt with immediately. Moreover, the architect will try to diffuse any tension in order to avoid protracted contractual disputes, and to ensure that all contractual matters are dealt with in the proper manner.

9.6 THE PRODUCTION PROCESS

The production of construction projects commences after a contractor's mobilisation and site set-up and, generally, the production process involves the effective use of resources (personnel, money, machine) and production techniques (methods, planning, organisation, motivation) to achieve the projects objectives. By and large, effective planning and organisation of the resources are prerequisites for the success of all construction projects, and the project team – especially the contractor – must ensure that these resources are managed efficiently to effect success in the production process. A common resource management creates the discipline of forward planning both the project and routine operations, and the most appropriate forward planning for the project is known as *short-term project planning*.

9.6.1 Short-term project planning

The execution of a construction project involves the combined efforts of a group of people with a view to achieving set goals against a timescale in the ever-changing

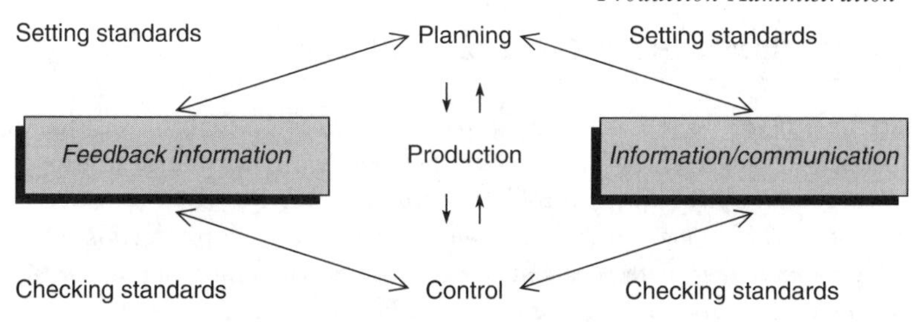

Fig. 9.3 Changing continuous activity in construction procurement

continuous activity of planning, production, control and planning again when there is deviation from standard (see Figure 9.3). Short-term planning encompasses all planning and programming activities undertaken after the production of the master programme and during the progress of the works.

Short-term planning enables master programmes to be broken down into greater detail for quantification and reflects the construction information and resources available for utilisation at any time during production. Hence, during the progress of the works, the site manager systematically plans and programmes each individual item of work on the master programme before its execution. Also, depending on the complexity of the construction project or a particular stage of its execution, short-term planning may take place daily, weekly or monthly. This is an indication that a construction project's planning is a continuous process which involves sub-contractors and trades supervisors who directly supervise the construction activities on site. Furthermore, this systematic and analytical pattern of the thinking process is beneficial to the contractor's site management, for the following reasons:

- It keeps the master programme under review and this leads to the establishment of reliable estimated dates for completion of activities and effective control over site operations.
- Site managers are obliged to think ahead and are therefore able to foresee immediate requirements and rearrange priorities in the light of the actual position of events.
- It reveals the information required at all times. This enables a steady flow of work to be arranged to reduce erratic work loading or non-productive time, which is the result of lack of production information.
- It enables site managers to learn from planning or production mistakes and, therefore, are able to institute corrective measures.
- It facilitates the introduction of imposed production techniques in the light of fuller information or investigation.
- It provides trades supervisors with a definite work programme and, hence, enables them to assess labour, plant and material requirements in advance.
- It enables interrelated problems to be identified well in advance as production activities are split up in minute detail.

9.7 PRODUCTION SUPERVISION AND CONTROL

During the progress of the works, the project team exercise control over site performance to ensure that what they have planned and programmed to happen is going according to plan. The control is a continuous activity which occurs throughout the production phase from commencement of site production through to final commissioning. Effective control entails a regular comparison of actual progress or performance against the predetermined programme and, where necessary, the introduction of appropriate corrective measures to achieve the desired objective. The production control is exercised by both the architect and the contractor, and the method and level of control depend upon a variety of factors such as the type, size, value and complexity of the project and the rate of production. Generally, the control process is to measure actual performance against the planned performance and the criteria of evaluation may be quantity, quality standard, time, cost, planning restrictions and so forth.

9.7.1 Architect's control process

Traditionally, as a project team leader, the architect's function in the control of the production process is fundamental to the successful completion of the project. However, his or her control at this stage of the project is restricted to the control of quality and rate of production and not the control of the technicalities of production. This exercise is to ensure that the building is constructed to the specified quality and time parameters. For this reason, the architect must communicate his or her requirements to the contractor effectively at all times during the life of the project. This may be achieved through the following mechanisms:

Site meetings

The architect may arrange a meeting with the contractor specifically in order to clarify design detail or help solve urgent construction problems which cannot wait until the monthly site meeting.

Routine site inspections

This entails a routine and regular inspection of the works in progress. The frequency of these inspections and the time taken by each inspection depends on the size and complexity of the project and/or the rate of production. Whenever the architect wishes to give production instructions during the site inspections, it is given to a general supervisor or the site manager and not direct to any site operative. An instruction given in this manner is known as a *site/verbal instruction* and the contractor must receive confirmation from the architect in the form of a formal written instruction. Following the proper channels of communication is important to maintain the smooth running of the project, and the architect ensures that this procedure is observed.

Records and reports

Where a clerk of works is employed on the construction project, the architect studies and comments on the weekly reports received from that person. This report covers several items, such as the number of operatives employed daily on the project, the weather conditions and its effect on the project, principal deliveries of materials and visitors to the site. These events are recorded in the clerk of work's site diary which keeps the architect fully informed of the day-to-day site activities. In addition, the clerk of works' site diary becomes a useful reference point for the resolution of contractual disputes.

Samples and testing

The architect may call for samples of various materials/components required for use on the construction project for his or her comments and/or approval. The architect can, if required, ask for sample panels to be prepared on site to facilitate the judgement of the visual effect of the specified materials in their intended position. In addition, he or she may instruct the testing of some materials such as concrete and bricks to determine their compressive strength and so forth.

Correspondence

Generally, the architect writes letters to clarify points in contention, to give general information or to confirm conversations that took place, for instance, during site inspection or a site meeting with the contractor.

Site instructions

The architect, during a site inspection, may issue site instruction(s) to rectify, change or clarify design detail, omit, add to or condemn an item of work.

Architect's instructions

The architect may issue a formal instruction to confirm a site instruction, rectify, change or clarify design detail, omit, add or condemn an item of work.

Drawn information

The architect may issue drawings, and drawings by other consultants, containing revised, amplified or clarified details of how certain items of work should be executed.

Defective work

Prior to handover, the architect carries out a final inspection of the works and lists those items which he or she is dissatisfied with by reason of non-conformity with the specification. The architect usually issues a defects list, often referred to as a *snagging list*, for immediate rectification by the contractor.

Contract administration meetings

These are the monthly site meetings of the project team. In these meetings, the contractor submits a progress report for the architect's perusal and comments. Apart from this, these monthly meetings enable the architect to assess the contractor's construction information requirements and attend to them quickly in order to avoid production delays and costly construction disputes. These meetings also assist the architect in his or her coordinating role.

9.7.2 Architect's coordination role

Generally, coordinating involves ensuring that all groups and persons work efficiently and economically, in harmony towards a given common objective. As a continuing process, it requires an efficient network of communication, both verbal and written. In a construction project, the common given objective is the solution of the problem posed by a client's stated requirements and hence, there must be a clear understanding of why the project team was formed and their general purpose. The architect performs the coordination function by creating a forum for the expression of conflicts, exchange of ideas, solution of problems and so on. This move establishes a direct contact between the persons immediately concerned and, at the same time, encourages the communication process on which the success of the project depends. For this reason, during the production phase of the project, the architect uses meetings as an instrument for ensuring that information is produced, recorded, disseminated and exchanged efficiently and that construction delays and disputes are avoided or minimised. These formal meetings, known as project team meetings or site meetings, should not be confused with either the architect's site inspection (and a short meeting this may bring about) or the numerous meetings the contractor may have with others, such as sub-contractors and suppliers. Initially, this meeting is held monthly but the frequency may be altered when required. Normally, the following procedure is followed in project team meetings.

Notice to attend meeting

The architect sends a reminder letter out to all those who are expected to attend the meeting. The reminder letter usually contains information of the date, place and time of the meeting and, also, enquiries of any specific requests for inclusion on the agenda.

Agenda for meeting

The architect prepares the agenda which is sent out shortly after the notification letter. The agenda should be specifically designed to meet objectives of the meeting and should also include all items requested by persons attending the meeting. An agenda of a meeting varies from project to project; however, the following list indicates some of the items of a typical project team meeting.

1. *Introduction:* All present at the first project team meeting introduce themselves – their names, positions and the company they represent. This formality is repeated whenever a newcomer attends any of the project team meetings.
2. *Apologies for absence:* Under this item, any person who was invited to the meeting but was unable to attend and has sent an apology for non-attendance is announced and recorded.
3. *Acceptance of minutes of the previous meeting:* The minutes of the previous meeting should be accepted as a true record of what was discussed. However, in the event of errors or inaccuracies, these should be rectified and formally recorded before acceptance.
4. *Matters arising from the minutes:* Any unresolved matters in the previous meeting, which are not included elsewhere on the agenda (and which require discussion and solution in the current meeting), are discussed.
5. *Main contractor's progress report:* The main contractor will give an account of progress achieved so far (i.e. since the last project meeting) by advising members of his or her progress report. In the process, the causes of any deficiencies or delays in the actual progress, compared to that programmed, will be explained for the architect's comments. The main contractor will also give an account of the position of various resources on site (e.g. number of trades and/or operatives, number and types of plant and quantity of specific materials on site and when more are expected).
6. *Nominated sub-contractor/supplier's report:* The nominated sub-contractor/ suppliers, if present, will give an account of the percentage of their work executed or supplies made so far and highlight problems (if any) or information requirements.
7. *Architect's/clerk of works' report:*

   (a) The architect's report generally covers the contractor's overall progress on the project, quality of workmanship and condemned work (if any). He or she will comment and record any sub-standard materials or use of incorrect construction methods noted and follow this up later by the issue of appropriate instructions.
   (b) The clerk of works' report normally covers items such as quality of labour and workmanship; construction methods; site safety, cleanliness, and so on; list of defects and confirmation of percentage of work completed.

8. *Consultant's report:*

   (a) The quantity surveyor may report on the progress made on matters such as evaluation of contractual claims, valuation of variations, re-measurement of any provisionally measured work, and the like.
   (b) The structural engineer may report on the progress made so far on the solution of a structural problem and on any outstanding matters awaiting further information/investigation.

(c) The services engineer may report on the progress made in solving the problem of coordinating the routes and location for various services element/equipments within the building.

9. *Sundry matters:* The items that are normally discussed under this heading include:

   (a) General communication problems – if there is any problem regarding communication among the project team, it will be brought up by a concerned member for discussion.
   (b) Information required – the contractor will indicate what information is required, when it is required and from whom, to comply with the programme.
   (c) Construction problems – if the contractor is experiencing any construction problems, these will be disclosed for discussion.

10. *Any other urgent business:* Any important matter not discussed in any of the above list of items is brought up for discussion.
11. *Date of next meeting:* The date for the next meeting is set in advance to suit all present.

It is important that the minutes taken for all site meetings are distributed to all those present, those who could not attend and any other relevant people on a circulation list before the date of the next site meeting.

9.8 CONTRACTOR'S CONTROL PROCESS

Contractor's control over the execution of the projects takes several forms and generally its extent depends on the size, nature and complexity of the particular construction project, contractor's skill and organisational and control policies. However, whatever the determining factors, the objectives and principles of control are the same; that is, to measure progress or result against a predetermined standard. The planning and programming establish the standard against which site progress will be measured and, to achieve effective control, the contractor ensures that resources for the construction of the project are efficiently combined and the production constantly monitored (see Figure 9.4). The monitoring cycle starts with measuring the actual output and comparing it against the planned performance. When this exercise reveals deviation, the causes of the deviation are analysed and corrective measures devised and implemented to correct the variance. The cycle is repeated by measuring the revised performance and comparing it to the standard. The process is repeated and the appropriate measures adopted until the planned performance is achieved. The desired result of this control is judged in quantity, quality and safety aspects of production. Also, the economy of production depends on the establishment of effective site control to achieve proper allocation of well-defined tasks, setting of targets and making sure that the targets are not lost by reason of:

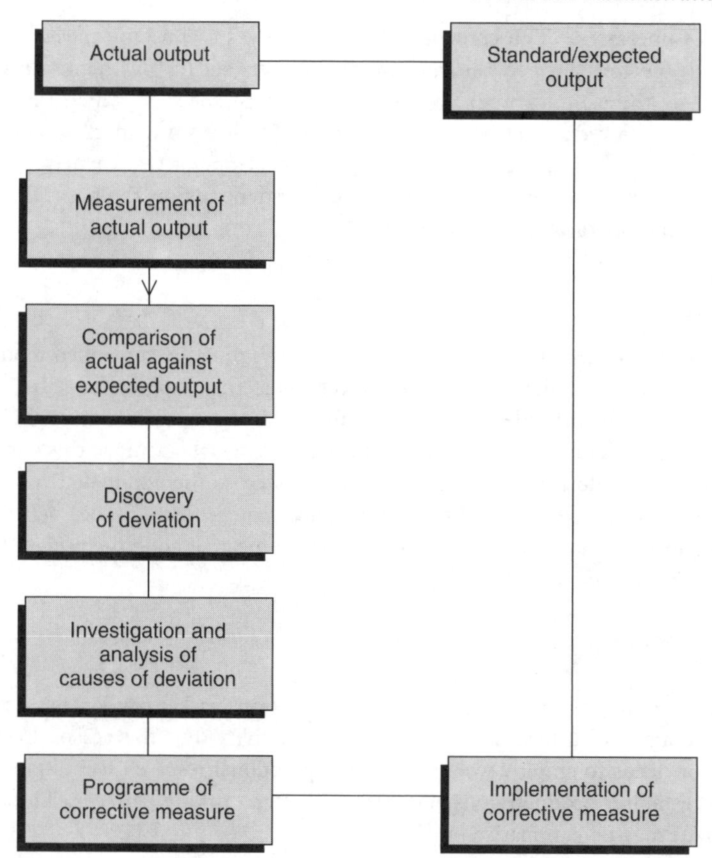

Fig. 9.4 The control process

- Poor performance.
- Use of unplanned or unspecified production method.
- Use of untrained or inefficient operatives.

9.8.1 Areas of control

The primary control areas are usually developed in the project planning phase and they include: labour, sub-contractors, productivity, materials, plant and quality standards

Labour

The amount of labour required for each item of work on a project can be assessed from the contractor's programme and the site manager, in conjunction with the trades supervisors, is responsible for the engagement, direction and control of operatives. The control of labour involves a constant check on operatives' performance in comparison with targets set for payment. This is achieved by the use of a weekly labour employment sheet on which the operatives' achieved target has been recorded. The record of performance, compared to those set, facilitates the calculation of operatives' earned bonus and weekly earnings. The success of this procedure depends on the accuracy of recording the operatives' daily output.

Sub-contractors

A sub-contractor, whether domestic or labour only, must be controlled to give increased output in order to meet the project's objective. Nominated sub-contractors, on the other hand, must be influenced to cooperate and do their best not to frustrate the efforts of operatives in other trades. To achieve this aim, all relevant project information must be speedily passed to the nominated sub-contractors concerned. Also, all their attendance requirements should be provided quickly and they should be encouraged to communicate among themselves to facilitate the solution of site problems.

Productivity

Increased competition for a share of the construction market necessitates economy of production, which means high levels of productivity. For this reason the contractor needs to employ work study and operational research techniques in order to establish productivity standards and better working patterns. These techniques may be summarised as follows:

1. *Work study:* The work study technique is composed of method study and work measurement. Method study seeks to improve production methods by the adoption and application of:

   (a) Better site layout and construction methods.
   (b) Effective equipment and handling techniques.
   (c) Better working conditions.

   and work measurement seeks to establish the basis of comparison and control to facilitate effective programming by undertaking:

   (a) Assessment of performance of labour and plant.
   (b) Application of costs to expected performance.

Materials

The basis of material control is a materials schedule which itemises all the materials necessary to complete a project. In addition, to ensure that the right quantity of materials is purchased, the contractor's buyers should derive the required quantities of materials by measuring from the architect's drawings. Apart from the main objective of establishing the right quantities of materials, this exercise serves as a check against the amount of materials contained in the bills of quantities.

The next approach to material control is the preparation of a materials checklist for delivery, maintenance of reliable delivery record and inspection and checking quantity/quality of key materials. There must also be control over waste during use of material for the works. Moreover, as a check of how successful the implementation of material control was on a project, a final material reconciliation should be carried out at the end of the project. This exercise involves comparison of total materials purchased for the project, less those retained or resold against actual quantities incorporated into the works.

Plant

The employment of plant in construction work reduces the amount of physical energy required in the execution of some operations. It also accelerates the pace of the construction process as it facilitates the execution of some mundane work of lifting heavy materials and transporting. As the contractor requires the use of plant in the execution of some sections of the project, it is therefore essential that he or she keeps records of all the up-to-date plants for construction works.

Plant hire or ownership is subject to a rate for its hire. When a plant department supplies its own company, it does so on the basis that the contract account will be charged on a weekly or hourly rate. This hire rate can be expensive and the contractor should control its effective utilisation once an item of plant is acquired. Economic utilisation of plant depends on:

- Sufficiency of work available to keep the plant in maximum working capacity during the period of acquisition.
- Experienced and skilled plant operators.
- Matching capacities of all plants to avoid under-utilisation due to mis-match of plant capacities.
- Careful planning and continuous programming of work to reduce wasteful idle standing time.
- Good maintenance of the plant.
- Selecting correct plant for the work.
- State of health and motivation of plant operators.

The above is an indication that, in the plant control process, the contractor adheres to the factors listed to ensure that plant time is not wasted. In addition, the plant control process extends to the recording of the delivery and installation of plant on site, maintenance, servicing and repairs. Equally important is the necessity to return all items of hired plant as soon as it becomes redundant on site.

Quality standards

During the design phase, it is the responsibility of the designer to ensure that quality is attained to the satisfaction of the client. Therefore, matters concerning quality are set out in the conditions of agreement between the client and the contractor. Also, quality standards which the contractor is required to achieve are specified in the contract documentation. The purpose of setting the standard is to produce a practical, factual and measurable limit as an object.

- Employment of skilled operatives with technological know how and who will exercise care in the execution of the works.
- Use of specified materials and components which are adequately tested, carefully stored and protected.
- Effective planning and programming of the works and controlled use of resources (i.e. plant, materials and operatives).
- Periodic inspection by site management and communicating inspection result to operatives.
- Adequate and effective protection of completed work.

At the production phase, the achievement of the quality standard specified becomes one of the pressing requirements of the construction project. Although several project participants with diverse specific interest and terms of reference are responsible for inspection, testing and approving various matters concerning quality, the contractor is responsible for the production of a quality product and, for this reason, has a duty to inspect and maintain a high standard of work.

SUMMARY

The coordination and control processes outlined above are essential as they lead to successful completion of construction projects. This function therefore requires constant inspection, communication, testing, planning, programming and reprogramming to achieve the set standard of output and quality. Sound planning is also the cornerstone of effective production management and the contractor ensures that he or she gets the project started on a solid footing and hence prepares adequately before the commencement of production on site.

FURTHER READING

The Aqua Group 1982 *Contract administration for architects and quantity surveyors.* Granada.

Calvert, R.L. 1976 *Introduction to building management.* Newness–Butterworth.

CIOB 1984 *Programmes in construction – a guide to good practice.* The Chartered Institute of Building.

Cooke, B. 1981 *Contract planning and contractual procedures.* The Macmillan Press.

Cornick, T.; Osbon, K. 1994 A study of the contractor's quantity surveying practice during the construction process. *Construction Management and Economics,* **12**, pp. 107–11.

Cottrel, G.P. 1979/80 The builders quantity surveyor. *CIOB Surveying Information Service, No. 1.*

Davies, W.H. 1982 *Construction site production checkbook 4.* Butterworth Scientific.

Forster, G. 1978 *Building organisation and procedures.* Longman.

Gray, C. 1981 Management and the construction process. *Building Technology and Management,* March, pp. 18–21.

Harper, D.R. 1990 *Building – the process and the product.* The Chartered Institute of Building.

Horner, R.M.W. 1982 Productivity, the key to control. *CIOB Technical Information Service, No. 6.*

Newlove, J. 1979 The issue of construction information. *Building Technology and Management*, July/August, pp. 2–5.

Pilcher, R. 1975 *Principles of construction management.* McGraw-Hill.

CHAPTER 10

COST CONTROL AND PAYMENT

10.1 INTRODUCTION

Cost control may be described as a process whereby a construction project cost is intentionally managed in every way possible so that:

- The client's authorised cost of a construction project (i.e. the contract sum) is not exceeded.
- The contractor does not lose profit (i.e. production cost exceeding the amount payable under the contract terms).

At the project's production phase, it is the duty of the client's professional advisers to constrain the actual project cost to conform to the planned project expenditure. At the same time, it is the duty of the contractor's quantity surveyor to oversee that the contractor produces the project within its original tender sum. To achieve these objectives, the following elements must prevail:

- Planning – formulate production cost plan.
- Monitoring – continuous comparison of actual with planned.
- Action – rectification of divergences from planned.

10.1.1 Managing cost

The responsibility for managing the authorised project cost is shared between the client's and the contractor's quantity surveyors, but for varied objectives. The client's quantity surveyor is under an obligation to the client, by the terms of engagement, to manage the cost (i.e. the contract sum) of the construction project effectively and, moreover, to do so impartially. Conversely, the contractor's quantity surveyor is more commercially minded and, for this reason, his or her role is to manage the project for financial success for his or her employer (the building contractor). The various functions required for managing the project cost at post-contract phase include the following:

- Establishment of budgetary control procedures before commencement of production.

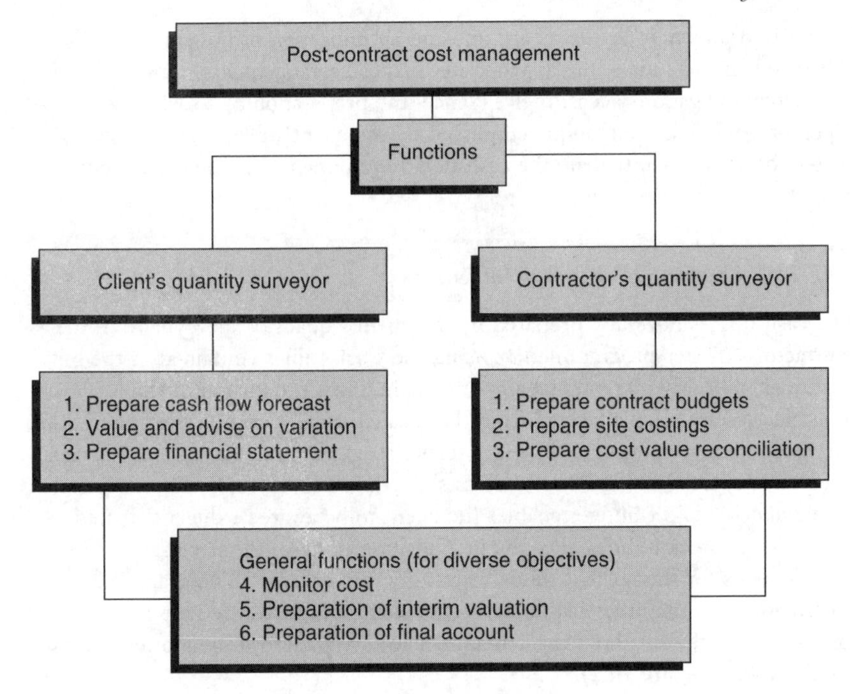

Fig. 10.1 Post-contract cost management functions

- Report on financial implications of proposed variations before the issue of formal instruction.
- Preparation of interim valuations to effect progress payment.
- Preparation of periodic cost reports (e.g. financial statement, cost value reconciliation).
- Regular comparison of planned cost or cash flow against the actual.

The various functions performed by the client's and the contractor's quantity surveyors are shown in Figure 10.1.

10.2 CLIENT'S QUANTITY SURVEYOR

The post-contract cost management functions of the client's quantity surveyor are summarised below (see Figure 10.1).

10.2.1 Cash flow forecast

Before or immediately after commencement of site production, the client does not merely want the construction of the development project to evolve without a clear knowledge of the financial outlay. For this reason, the professional advisers provide information on when payments are due and an indication of how much is due, so that the client may make the appropriate financial arrangements to meet the contractual obligations accordingly. Hence, the preparation of a cash flow or expenditure flow is a technique employed by the client's professional advisers to enable the client to anticipate the cash flow requirements for the construction project.

10.2.2 Preparation of cash flow forecast

The cash flow is normally prepared by the client's quantity surveyor from the contractor's master programme. By using the work values contained in the bills of quantities, monthly incremental amounts including a proportion of the preliminaries are calculated and from these calculations a cumulative expenditure curve can be plotted against time (see Table 10.1). A curve thus plotted is usually referred to as an *S Curve* because of its shape (see Figure 10.2). The preparation of the tabulated project budget enables the client to be aware of the anticipated expenditure forecast and, hence, be in a position to make the necessary arrangements for the needed funds when they become due. The graphical presentation, on the other hand, shows the client's expenditure pattern plotted against time and also allows the actual expenditure pattern to be plotted against the predicted (see Figure 10.2).

The graph shows the duration of production (i.e. commencement and completion dates) and table of expenditure. The expenditure curve is shown representing a smooth start, accelerating during the middle third, and then gradually tailing off during the end of the project. Against this same graph monthly interim valuations can be plotted (see Figure 10.5 on page 194).

10.2.3 Optimum expenditure pattern

As noted previously, the contractor's master programme represents the planned stages of construction production. But, at the same time, it is an indication of how the client is expected to fund his or her construction project. The client's professional advisers also look for an optimum utilisation of the client's funds and, for this reason, the client's funds must not be tied up in unproductive uses such as construction work in progress that is not able to generate any revenue or income until it is completed. Hence, again the client's quantity surveyor ensures that the contractor's programme is realistic in terms of duration and the level of financial charges it attracts. Therefore, any construction programme which shows heavy initial expenditure, as curve A in Figure 10.3, is questionable. This curve may be described as a front-loaded project programme; it shows a rapid project start-up and an even more gradual than normal phase out at the end. At the same time, any construction programme which shows heavy expenditure towards the end of the project, as curve B in Figure 10.3, is risky as a result of inflationary pressures and uncertainty surrounding the projects completion on a target date. This is a back-loaded project programme; it indicates a more relaxed start and a very steep finish

Table 10.1 Tabulated presentation of project budget

Contract: Upper Dagnall Street Development Project																			
		Time in months																	
Operation	Cost (£)	1	2	3	4	5	6	7	8	9	10	11	12	13	14	15	16	17	18
Site preparation	50,000	40	10																
Sub-structure	410,000	50	80	120	100	60													
Drainage	54,000		14	20	20														
RC frame	498,000				60	130	130	130	48										
Masonry	40,000							15	25										
Cladding	162,000								50	50	50	12							
Services	370,000						60	60	60	60	60	60	10						
Roof covering	110,000									30	30	30	20						
Partitions	160,000								10	30	30	30	30						
Carpentry/joinery	227,000							17	30			30	30	30	30	30			
Internal finishings	47,000													2	15	15	15		
Glazing	90,000														1	9	2		
Painting	12,000						10	10	20	35	10					20	22	26	
External works	202,000												27	30	13	10	2	10	15
Preliminaries	296,000	17	17	17	17	17	17	17	17	17	17	17	17	17	17	17	17	17	17
Monthly totals	**2,728,000**	**107**	**121**	**157**	**197**	**207**	**217**	**249**	**270**	**222**	**197**	**179**	**134**	**109**	**97**	**92**	**87**	**62**	**24**
Cumulative totals	**£0000**	**107**	**228**	**385**	**582**	**789**	**1006**	**1255**	**1525**	**1747**	**1944**	**2123**	**2257**	**2366**	**2463**	**2555**	**2642**	**2704**	**2728**

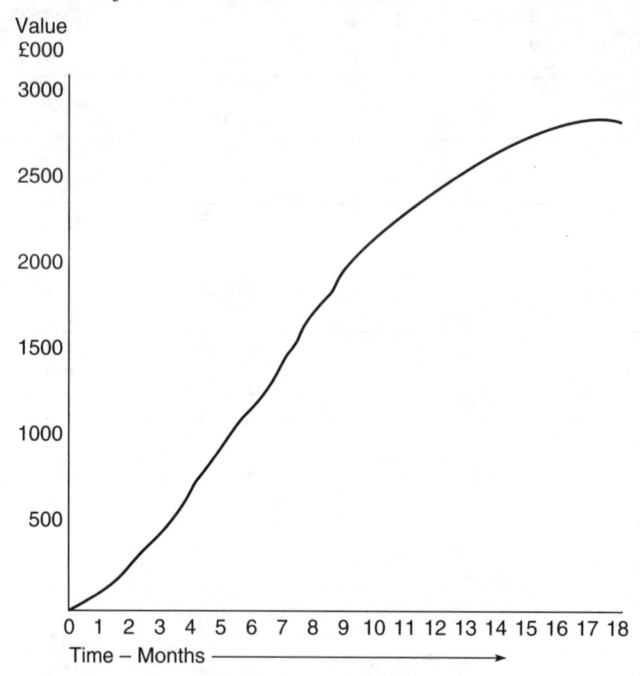

Fig. 10.2 Graphical presentation of project budget

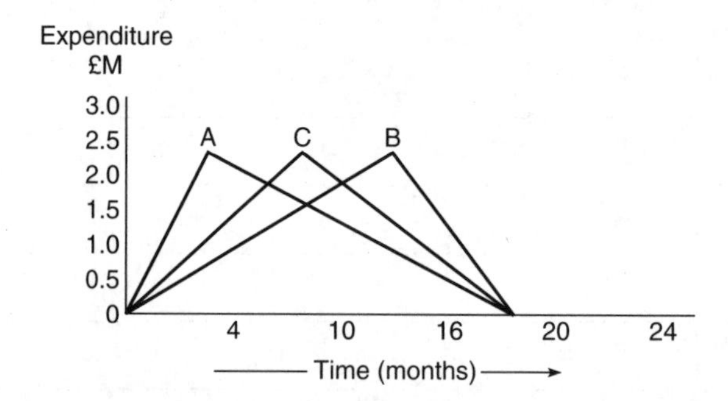

Fig. 10.3 Simplified idealised incremental expenditures

slope on the curve. The steep finish is likely to lead to such problems as inefficient use of resources and overrunning of the project budget and time. The inefficiency is the result of allocating too much resources to the few remaining tasks programmed for fast completion in order to meet a target date.

The client's quantity surveyor is also expected to advise on implementation and enforcement of a realistic construction programme which balances expenditure and progress and results in an expenditure pattern denoted by curve C in Figure 10.4 rather than curve A or B. This is an optimum production programme which safeguards the client from high initial expenditure (which attracts more interest

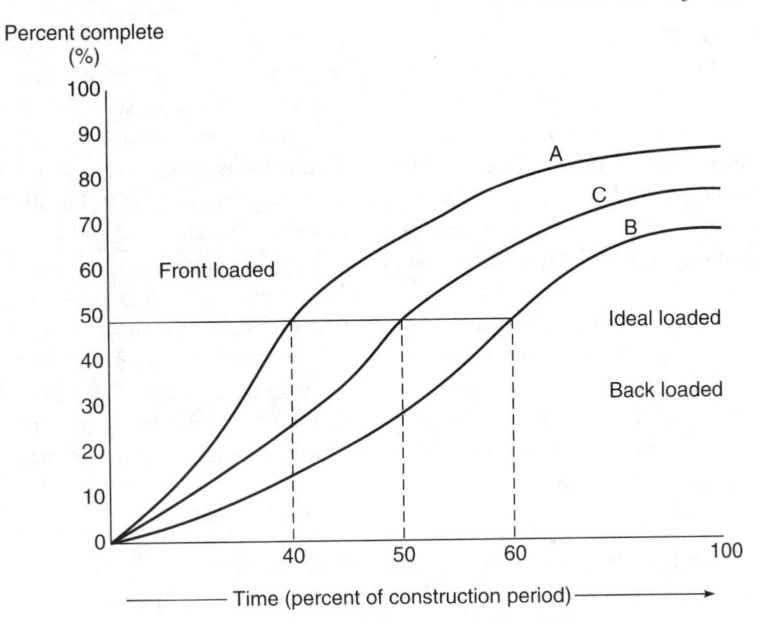

Fig. 10.4 Percent completed vs scheduld time

charges as in curve A), inflationary pressures and uncertainty of timely completion (as in curve B). As opposed to curve A (where 50 per cent of the project is completed in 40 per cent of the time) and curve B (where 30 per cent of the project is completed in 50 per cent of the time), in curve C, 50 per cent of the project is completed in 50 per cent of the project time. This is an ideal project programme which is able to balance expenditure and progress and, hence, is worthy of emulation.

10.2.4 Valuation of variations

During the production phase of construction projects, variations do arise in one or more of the following situations:

- An architect/client's need or wish to vary sections of the original design and/or specification.
- Discovery of ambiguity in the contract documentation (e.g. discrepancy between drawings and bills of quantities or between architectural and engineering drawings) and consequent clarification.
- Discovery of an omission or error in the contract drawings, bills of quantities or specification.
- An action taken to meet statutory requirements under the orders of the building inspector, district surveyor or persons with a statutory right of control.
- Reduction of cost owing to financial constraints.

Traditionally, most construction contract conditions make provisions for adjustment of the contract sum due to variations to the design and/or specification of the work. The rationale behind these provisions is that, generally, these variations do alter (i.e. increase or decrease) the scope of the work and thus create a need for financial adjustment. Where no machinery for financial adjustment exists, variations can only be effected by fresh agreement between the client and the contractor. Normally, the valuation of variations requested and/or authorised by the architect is the responsibility of the client's quantity surveyor. It is that person's responsibility therefore, to ensure that this service is provided expeditiously to facilitate the provision of an effective cost management. However, where the objective of the valuation is for the provision of cost control, the architect should, in the first instance, ask the quantity surveyor to value the proposed variations and provide information on the financial implications. With the full knowledge of the financial effects of the proposed variations, the architect may take one or more of the following courses of action:

- Where no harm will occur from non-action, the architect may change his or her mind and leave the design and/or specification intact.
- The architect may ask the quantity surveyor to carry out cost studies on alternative design solutions and/or advise on the most economical proposition.
- The architect may accept the financial effect of the variation and issue formal instructions.

The above practice enables the architect to consider the cost effect of an instruction before an expenditure is committed; but the absence of the above procedure (from the client's quantity surveyor's point of view) reduces the objective of valuation of variations to a cost-monitoring and reporting exercise, rather than cost control. If the former is the case, as the project progresses, the quantity surveyor makes periodic assessment of the value variations before the architect's instruction is issued. To be effective also, the process needs extremely close liaison between the client's professional advisers.

10.2.5 Financial statements

A financial statement is a periodic cost report prepared by the client's quantity surveyor to give the architect and the client an up-to-date financial status of the construction project. This report is composed of the total value of variations to date, cost of unconfirmed variations, fluctuations and possible contractor's claims for direct loss and/or expense. Although the quantity surveyor has no direct control over the instruction of variations during the project production phase, he or she can predict their financial effect and thus make the client aware of future financial commitments. The information contained in the report is normally intended for the sole use of the client. For this reason, at times, it necessitates the inclusion of a disclaimer caution in the event of a contractor utilising it to justify a contractual claim.

Example 10.1 gives a sample of the financial statement.

Example 10.1 Financial statement

FINANCIAL STATEMENT NO 5

Contract:
Upper Dagnall Street Development Project
Contract period: 78 Weeks
Date for practical completion:

Current Financial Statement
As at: (Date)
Extension: Nil Weeks

		£
1. Authorised commitment		
(a) Original contract sum		2,728,043
(b) Approved supplementaries		Nil
	Current total approval	**£2,728,043**

2. Variations and adjustments

	Omissions	*Additions*
(c) Adjustment of provisional sums	25,000	17,500
(d) Adjustment of PC sums	75,000	67,500
(e) Provisional quantities	17,360	19,425
(f) Architect's instructions (Nos. 1 to 35)	54,250	63,900
(g) Variations awaiting architect's confirmation	1,250	2,790
(h) Fluctuations in labour and materials	Nil	2,250
(j) Anticipated variations	Nil	Nil
(k) Contractual claims awaiting evaluation	Nil	3,250
	£172,860	**£176,615**

Net omissions/additions*	£3,755
Total approval to-date	£2,728,043
Anticipated total expenditure	£2,731,798
Less Total approval	£2,728,043
Total overspent/underspent*	£3,755

(* *delete as appropriate*)

10.3 CONTRACTOR'S QUANTITY SURVEYOR

Like the client's quantity surveyor in the construction process, the contractor's quantity surveyor performs the functions detailed below in the management of the contractor's cost of executing the project.

10.3.1 Contract budget

A budget is a forecast or plan for the future against which to monitor and measure performance. In the construction industry, a budget is a financial plan for a construction project as a whole and is used to determine cash flow

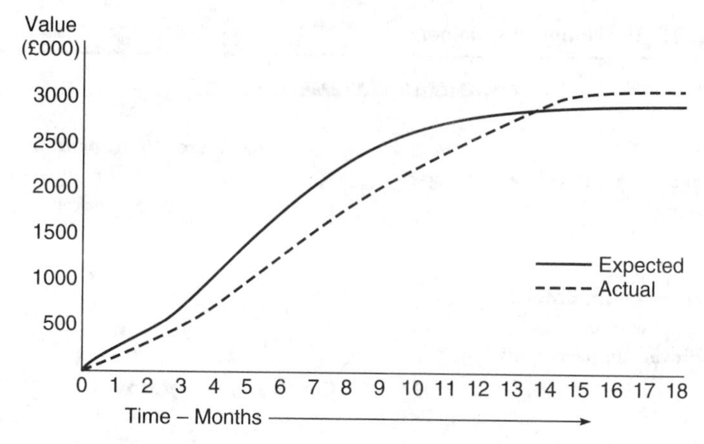

Fig. 10.5 Actual interim payments against expected

requirement for the duration of a project. In addition, a budget serves as a yardstick against which actual production of a project can be measured. Therefore, as a means of forecasting expenditure/cost, interim valuations and income, the contractor's quantity surveyor produces various financial budgets which permit contract performance to be monitored and controlled. The financial budgets which the contractor's quantity surveyor may prepare during the pre-contract stage include:

- Cumulative value/time budget – used for obtaining forecast of interim valuations.
- Cost/time budget – used for monitoring actual site progress.
- Income/time budget – used for forecasting cash flow requirements.

Each of the above financial budgets is produced from a project master programme and presented in graph or chart form. The most popular and simple of all cost curves is the S curve for cumulative expenditure/value plotted against time (see Figure 10.5).

10.3.2 Site costing

During the progress of the works, the contractor's production managers require periodic cost information in order to manage the project efficiently, and production cost is among the cost information that management needs. It is, therefore, the responsibility of the contractor's quantity surveyor to carry out site cost checks and provide management with the required information. In effect, site costing provides management with feedback of the efficiency of past decisions and thus enables them to change course and, if necessary, introduce appropriate corrective measures.

Currently, several site-costing systems are being adopted by building contractors for construction production and management, but any good costing system should have the following qualities:

- Facilitating the establishment of plans, methods and achievable targets for construction works.
- Possession of the mechanism for comparing cost targets with actual performance.
- Facilitating the assessment of resource requirements in line with methods and achievable targets.
- Ability to provide an early warning of shortfall on productivity.
- Assisting in the determination of profit/loss on cost in an accounting period.
- Capability of establishing methods of control and valuing work in progress.
- Provision of information which assists management in financial decision making.

In order to provide an efficient site costing, the contractor's quantity surveyor needs an appreciation of elements of cost and site-costing techniques.

Elements of cost

The cost of a unit of output per period can be regarded as being built up of a number of elements of cost, and these may be summarised as follows:

1. *Direct cost:* Direct cost comprises the cost of materials, plant, wages and all direct expenses. This cost varies in proportion to activity and therefore it is also known as a variable cost.
2. *Site overheads:* This is composed of overheads incurred in production (i.e. salaries and expenses on site staff). As far as a construction project is concerned, a site overhead is also known as a fixed cost because it remains unchanged despite changes in activity over a fixed time span.
3. *Head office overheads:* This consists of overheads incurred in managing a contractor's business. It includes top management costs, directors' fees, rents and the like. As this cost remains unchanged in the short run, despite changes in the contractor's business activity, it is also known as period fixed costs.
4. *Profit:* This is the difference between the total value and the total cost of production.

Site-costing techniques

Depending on the purpose for which management require the information on site costing, the contractor's quantity surveyor should be able to use one or more of the following costing techniques to provide the required site information.

1. *Marginal costing:* The technique of marginal costing is based on an assumption that, since period fixed costs are a constant amount, the production of one extra unit of output causes total costs to rise only by the variable cost (i.e. the marginal cost) of production. Similarly, total cost falls by the variable cost per unit for each reduction by one unit in the level of activity. To adopt this technique, the contractor's quantity surveyor should be able to identify and relate the project variable overheads and preliminaries and allocate their cost accurately to the production tasks or site activities. The technique is intended to highlight increases in variable costs for management's control and/or action.

2. *Absorption costing:* This is a costing technique which charges a proportion of all overheads to site activities on the assumption that overheads are contributed by all the production sections. In other words, this is a means of charging overheads to site activities, and in construction activities the overhead rate is normally charged on a percentage basis. This means that in every site production activity, a percentage covering overhead is added to the elements of cost (i.e. labour, plant and materials) in proportion to the contractor's annual turnover. Moreover, the percentage additions allocated to the site activities are based on the anticipated level of activity and, therefore, if there is a shortfall in the level of activity, the contractor will not be able to recover the overhead cost.
3. *Standard costing:* The standard costing technique involves estimating the costs of production activities and services. This method of costing provides management with information regarding deviations from a production plan. Therefore, it compares the actual cost of an activity with what it ought to cost and, hence, provides a basis for cost control. To adopt the technique, the contractor's quantity surveyor will need to carry out the following activities:

   (a) Determination of cost for each site activity.
   (b) Monitoring, calculating and recording actual costs incurred.
   (c) Periodic comparison of actual costs with the predetermined standards in order to ascertain deviations.
   (d) Investigating the reasons for deviations and informing management for corrective action where necessary.

   The standard costing technique covers both overheads and direct costs associated with construction production. For this reason, at times, it complicates the determination of target costs per unit of production for plant, labour, materials and daily site control.
4. *Contract costing:* Contract costing refers to a technique of costing employed where work is undertaken to the customer's special requirements and where the production of each order is of long duration. When contract costing is adopted in the construction production, a separate number is allocated to each project. Due to the magnitude of construction projects, in order to maintain adequate control, the contractor's quantity surveyor breaks down each construction project into a series of sub-order/sections and the summary of sub-order/sections represents the cost of project to date. The sub-order/section costs or targets can be the subject of control within the contractor's establishment and the production cost target of sub-order/sections can be compared with the actual cost of production at the completion of the sub-order/section.

10.3.3 Cost value reconciliation

The responsibility for the preparation of cost value reconciliations varies from company to company. However, the responsibility for its preparation at project level rests with the contractor's quantity surveyor. Being in control of the cost of site activities and the end results, he or she is in the best position to provide information on cost and value of production. Cost value reconciliation is an accounting practice for monitoring continuously the cost of a long-term project and work in progress in order to assess the projected final profit or loss. If projected final losses are identified early enough, corrective measures can be introduced immediately to arrest the situation. Although, construction companies have diverse policies for carrying out financial forecasts, reporting and reconciliation, ideally the cost value reconciliation exercise should be executed on a monthly basis (after preparation of interim valuation) to ensure a sound financial control.

The contractor's quantity surveyor normally produces the internal valuation (a typical one is shown in Example 10.2) to enable the company's accountant to carry out the essential function of matching the true value of work done to its associated cost. The reconciliation function involves comparison of the quantity surveyor's internal valuation against total cost elements (i.e. materials, plant, site staff and overheads, labour and sub-contractors) to determine whether a running construction project is in either a profit- or a loss-making position within a period under review. At the same time, the cash position of the project is examined by comparing total payment application against gross certificate value to reveal the total cash outstanding (including retention and claims income) on the project account.

Example 10.2 Contractor's internal valuation

INTERNAL VALUATION

Contract number: 7		Project: Upper Dagnall Street Development Project	
1. CERTIFICATE			
Add	(a) Work done	730,444	
	(b) Materials on site	127,000	
	(c) Nominations	28,166	
	(d) Statutory undertakers	7,929	
	(e) Fluctuations	10,691	
			904,230
2. OVER-VALUATIONS			
Deduct	(a) Materials not paid	71,370	
	(b) Sub-contractors not paid	66,993	
	(c) Other non-payments	-	
			138,636
3. ADJUSTMENTS			
Add	(a) Variations not agreed	6,421	
	(b) Sub-contractor's accounts not agreed	Nil	
			6,421

Example 10.2 *Continued.*

INTERNAL VALUATION			
Contract number: 7		PROJECT: Upper Dagnall Street Development Project	
4. PROVISIONS			
Deduct (a) Remedial works		1,000	
(b) Anticipated losses		4,500	
(c) Defects liability		Nil	
			5,500
5. INTERNAL VALUATIONS			766,515
6. TIME ANALYSIS (in weeks)			
Original contract period	78	Present weeks	30
Extension of time awarded	–	Weeks to complete	48
Revised contract period	78	Anticipated overall period	78

10.4 GENERAL QUANTITY SURVEYING FUNCTIONS

There are some quantity surveying functions which are carried out regularly by both the client's and the contractor's quantity surveyors. Generally, these functions are performed separately; however, there are some which can be performed jointly by both surveyors but for different objectives. While individual quantity surveyors vary considerably in their methods of working, when two surveyors work together, they are normally able to deliberate and agree contentious points of common interest expeditiously. Additionally, this working arrangement saves time and later disagreements as points of difference can be resolved quickly. Again, both quantity surveyors can adopt a working arrangement where one works and the other checks. The latter working arrangement (i.e. where both surveyors work successively) leads to the same expeditious production of results and information. Cost monitoring, payment and final accounts includes some of the quantity surveyors, general functions.

10.4.1 Cost monitoring

Generally and contractually, during the production phase, the client's quantity surveyor has no authority to issue instructions which would, in any way, affect the cost of the construction project. For this reason and strictly speaking, he or she is in no position to control cost. Rather, he or she monitors the project's cost constantly, prepares periodic assessment of the anticipated final cost and reports them to others (the architect and client) to enable them to take necessary measures in the control of cost.

The contractor's quantity surveyor, on the other hand, is unlikely to have any direct control of the cost of production. Rather he or she monitors the contractor's production cost by the maintenance of a regular record of true costs which can be compared with the recoverable value, to enable management to assess the potential or actual profit/loss.

Although the cost monitoring is of great concern to both quantity surveyors, it is for different objectives. From the view point of the contractor's quantity surveyor, his or her concern centres around the total value of the project and the total cost of resources in use and sub-contractors. The client's quantity surveyor, on the other hand, considers the payment the client makes to the contractor for executing the construction project and reports immediately to the architect whenever this payment is likely to exceed the client's cost limit. The cost monitoring exercises keep both the client and the contractor abreast of their respective current financial positions. It also enables them to take prompt corrective measures where the need arises.

10.4.2 Payment

Like all other contracts, the construction contracts contain provisions for full payment to be made to the contractor for executing the client's construction projects. But as the final product is usually large, expensive and takes sometime to produce and hence, the costs incurred cannot be reasonably borne by the contractor, the construction industry has a long established tradition of a system of interim payment. For this reason, interim payments during construction production are deemed necessary to afford the contractor a flow of cash with which to progress the works to completion.

Methods under which interim payments are assessed are payment by time, payment by element of the work completed and payment by remeasurement. Each of the above methods has its strengths and weaknesses but the first requirement of any interim payment system is that it should be fair to both the client and contractor according to their needs. Normally, the client does not want to pay for any sections of the work earlier than necessary. The contractor, on the other hand, wants to be paid as early as possible for sections of work satisfactorily completed to keep cash flowing in for the execution of the project at the pace required by the contract. When there is persistent shortfall in interim payments, this could starve the contractor of cash and, moreover, hold up the project by low output.

Payment relating to time

This is an interim payment method in which the contractor receives payment for the total number of hours spent on the project at each predetermined interval. This payment system is suitable for a cost reimbursement contract where the contractor submits a signed record of total hours spent by his or her operatives for the assessment of payment due. This method has the virtue of simplicity but it does not provide the contractor with an incentive to progress the project at the pace required by the contract.

Stage payments relating to elements of the work completed

This method of interim payment is sometimes referred to as *stage payment, milestones* or *payment schedule*. Stage payment is the term for an interim valuation carried out only when an agreed defined stage of the construction work has been completed instead of at agreed time intervals. This means that during the progress of the works the contractor receives payment as and when the construction of a defined stage is completed. This method is suitable for housing projects with repetitive identical stages, and when the method is recommended both the stages to make payment and the value of each stage have to be agreed prior to contract. For instance, in a housing project, each house may be divided into eight or nine agreed stages of substructure, external walls, floors, roof, doors and windows, plumbing and electrical, internal finishing and decorating, external works and preliminaries and the value of each stage agreed.

Under the stage payment arrangement, the contractor receives payment on completion of each stage and, therefore, first payment is due when the substructure is completed. This is followed by external walls and others when they are also completed. Therefore, on a housing project, for instance, the total valuation recommended to the architect for certification reflects the value of stages completed in addition to unfixed materials and components stored on or off site.

The advantage of this method is that it aids progress of the construction project and at the same time contains strong motivational elements, as the contractor has a precisely defined target and gets paid when that target is reached. The assessment for the interim payment and client's cash planning are also simplified as a result of the predictable work stages. Its drawbacks are that since payment is only made against a completed stage, there is a scope for dispute over whether or not the work encompassed in each stage has been fully completed. There is also the risk that completion of a stage may be delayed by others, which could mean no payment at all to the contractor.

Payment by remeasurement

Payment by remeasurement is the most common method of calculating interim payment. In this approach, the work completed is assessed at predetermined intervals and the contractor is paid a sum equivalent to the work judged completed by the client's quantity surveyor. This method of interim payment is cumbersome, extremely time consuming and generates high overhead costs. Under this method there is also uncertainty over the amount of interim payment. For instance, the contractor's quantity surveyor might claim a higher percentage (say 60 per cent) of work completed while the client's quantity surveyor might claim that a lower percentage (say 40 per cent) of the work has been completed, leading to a lower payment than anticipated.

Preparation of valuation for interim payment (remeasurement method)

As noted above, an interim valuation may be described as an accurate assessment of the value of work executed satisfactorily by the contractor. Under the remeasurement method, the process involves valuation of the total amount of work which has been properly executed since the commencement of production, and the materials/components on site and off site (where instructed for inclusion by the architect) on the date of the valuation. The purpose of valuation is, firstly, to make recommendation to the architect for the issue of an interim certificate and, secondly, to finance the contractor during the progress of the works.

All interim valuations, in principle, should take place only when the architect considers them to be necessary. However, in practice the architect relies on this valuation as a means of periodic progress value check and, therefore, it is prepared (monthly or shorter periods) throughout the life of the project.

Although it is the responsibility of the client's quantity surveyor to prepare the interim valuations, as a matter of convenience this is normally prepared jointly with the contractor's quantity surveyor. However, in order to save time, invariably the interim valuation is prepared by the contractor's quantity surveyor and submitted to the client's quantity surveyor during their site valuation meeting for his or her perusal and agreement. Moreover, the advantages accrued when both surveyors prepare the interim valuation together is that several points of difference can be resolved leading to the production of a quick and fair valuation.

The general procedure for preparation of interim valuations is that both quantity surveyors arrange and confirm the valuation meeting date and time before they meet on site. However, at times, interim valuation dates are agreed in advance (e.g. the last Friday or the 27th day of each month) at the project briefing meeting.

10.5 SURVEYORS' PREPARATORY WORK

10.5.1 Contractor's quantity surveyor's preparation

Prior to the valuation date, the contractor's quantity surveyor carries out some preparatory work to ensure the availability of all valuation information and by so doing avoids an under-valuation. Hence, the contractor's quantity surveyor carries out the following preparation before meeting his or her counterpart.

- Preparing interim valuations with all sub-contractors working on the project (groundworker, bricklayer, plasterer, painter) for work executed over the period under consideration.
- Advising all nominated sub-contractors and suppliers to submit their application for payment before the interim valuation date.
- Informing and advising all statutory authorities to submit their fees and charges (including materials and components) for work executed to date.
- Working through and listing all architect's instructions executed and their estimated values.
- Where an architect's instruction has been executed on a daywork basis, ensuring that the daywork is signed by the clerk of works and bears the architect's instruction origin or number, priced and ready for the client's quantity surveyor's decision whether or not to include it in the interim valuation.

- Preparing, listing and valuing all site instructions or contractor's verbal instructions which have been executed but awaiting the architect's formal written instructions.
- Calculating the estimated value of claim for direct loss and/or expense which the architect has been notified of but has yet to be prepared formally for submission.
- Remeasuring any executed items of work which have been provisionally measured in the bills of quantities.
- Preparing a list of materials and components on and off site.
- Preparing a list of construction problems such as discrepancies between drawings and the bills of quantities, under-measurements in the bills of quantities and so on for discussion and/or adjustment within the interim valuation.

10.5.2 Client's quantity surveyor's preparation

Before the valuation date, the client's quantity surveyor makes the following preparations before meeting the contractor's quantity surveyor:

- Going through the previous valuation and valuation notes (if any) to become familiarised with the details of the build-up of that interim valuation.
- Checking the minutes of site meetings and correspondence file to determine whether they contain any information which may influence the outcome of the current interim valuation.
- Examining the nominated sub-contractor's/supplier's application for payment and preparing a list of items worthy of inspection on site.
- Checking if all nominated sub-contractors/suppliers have received payment due to them in the previous interim valuation.
- Finalising the measurement and computation of any executed authorised variations and provisionally measured items in the bills of quantities for discussion and/or inclusion in the interim valuation.
- Examining the list of unfixed materials included in the previous valuation and noting materials which, by reason of their value/quantity, should reconcile with the list of unfixed materials on site for the current interim valuation.
- Checking the authenticity of daywork sheets received and noting those with excessive man-hours for clarification or adjustment before inclusion in the interim valuation.
- Finding out from the architect whether there are any defective construction works on site which should not be included in the interim valuation.
- Checking whether the client's authenticated VAT receipt for the previous interim valuation has been returned by the contractor.

10.6 INTERIM VALUATION

When both quantity surveyors meet on site, their first task is to tour the site, inspect the works and take notes as necessary of the proportion of work executed on each section, trade or element. The site notes should cover the main contractor's, sub-contractors' and statutory authorities' work, quantities of materials/components on site and any items of work worthy of discussion/attention.

On return to the site office, the value of each section, trade or element is assessed. Generally, the following items are normally considered in the preparation of interim valuations.

10.6.1 Main contractor's work

The main contractor's work comprises the following:

Measured work

Where priced bills of quantities are used, the value of work executed by the contractor can be determined from the measured work section. This process consists of going through the bills of quantities and extracting the value of completed and partially completed work items. The proportion of partially completed work items are assessed by visual inspection rather than by precise measurement.

Variations

Any authorised variations which have been executed and measured or valued/adjusted are included in the interim valuation.

Preliminaries

Preliminaries may be valued by the adoption of one of the following methods:

1. *Time related* (i.e. proportion per month/valuation period): Where the contractor priced preliminaries as a lump sum (and was not asked for a breakdown before the award of contract), the value of preliminaries is obtained by dividing the total preliminaries by the number of months in the contract period, thereby giving a constant amount for inclusion in each valuation, as in Example 10.3.

Example 10.3 Valuations of preliminaries: proportion per month method

Total preliminaries = £295,788.00

Contract period = 18 months

Calculation $= \frac{£295,788.00}{18} = £16,432.67$

If this method is adopted, the sum of £16,432.67 will be added to each monthly interim valuation. Moreover, this method assumes that the project will make a steady progress according to programme. However, where production delay occurs, it could result in a situation where the entire value of preliminaries is paid out before completion of the project. Should this happen, it will place the client in a disadvantageous position, especially where the contractor is unable to complete the project by reason of insolvency.

2. *Value related* (i.e. percentage of contract value): In this approach, the value of preliminaries is obtained by expressing preliminaries as a percentage of the contract sum (less PC and provisional sums) for addition to the contractor's monthly interim valuations, as in Example 10.4.

Example 10.4 Valuation of preliminaries: percentage of contract value method

	£	£
Contract sum		2,728,043.14
Less		
Preliminaries	295,788.00	370,788.00
Daywork	45,000.00	
Contingency	30,000.00	
		£2,357,255,14

	£	
Total Preliminaries	295,788.00	
Calculation	$\frac{295{,}788.00}{2{,}357{,}255.14} \times 100$	= 12.55%

Where this method is in use, 12.55 per cent of the total expenditure on completed works at every interim valuation period is taken as a fair representation of the preliminaries element and, hence, added to the interim valuation. For example, the proportion of preliminary element in a total expenditure of £240,000.00 on completed works will be 12.55 per cent of £240,000.00 = £30,120.00.

3. *Value related* (i.e. apportioning preliminaries in direct proportion to total expenditure on completed works): Under this method, the value of preliminaries is obtained by relating the current or percentage value of the measured work to the total preliminaries, as in Example 10.5.

Example 10.5 Valuation of preliminaries: apportionment in relation to total expenditure method

	£	£
Contract sum		2,728,043.14
Less		
Preliminaries	295,788.00	370,788.00
Daywork	45,000.00	
Contingency	30,000.00	
		£2,357,255,14
Total expenditure on completed works (say)		502,000.00
Total preliminaries		295,788.00
Calculation	$\frac{502,000.00}{2,357,255.14} \times 295,788.00 = £30,115.16$	

If this method is adopted, the above calculation will need to be carried out in each interim valuation period to obtain a proportion of preliminaries for inclusion in the valuation.

4. *Combination of event, time and value related:* In this approach, the preliminaries are broken down into event-, time- and value- related costs. Each is explained briefly as follows:

   (a) An event-related cost refers to preliminary items such as setting out and final cleaning of a construction site which are due on the occurrence of the operation or event. What is more, a preliminary event-related item can occur either at the beginning or at the end of a construction project.
   (b) Value-related cost refers to preliminary items such as water and insurance for the works whose values are related to the total value of the project.
   (c) Time-related cost refers to preliminary items such as supervision and hire charges which are proportional to the contract period.
   (d) Combination of event- and time-related costs refer to preliminary items such as site huts and standing scaffolding whose cost are made up of erection and removal (event-related cost), and hire and maintenance (time-related cost charges). If this approach is adopted, there will be a need for the client's quantity surveyor to analyse the preliminaries bills into the various constituents of cost and convert them into payment, as shown in Example 10.6.

Example 10.6 Valuation of preliminaries: event, time and value related (Breakdown of preliminaries into various constituents of cost methods)

1. *Data*

   Total preliminaries £295,788.00

   Contract period 18 months

   Valuation number 2

2. *Calculations*

Preliminary items	Value £		Event (initial) £		Time (related) £		Event (final) £
Supervision	-		-		4,000		-
Plant	-		4,500		2,100		1,000
Insurance	2,500		-		-		-
Site hut	-		3,500		100		750
Water	-		850		60		-
Drying out	-		-		300		-
Electricity	-		1,000		100		300
Disposal	-		-		75		-
Site transport	-		-		150		-
Temporary road	-		4,000		50		-
Health and welfare	-		100		1,500		-
Telephone	-		300		100		-
Site security	-		-		250		-
Sundry (assumed)	-		10,000		6,000		858
Totals	2,500	+	24,250	+	(14785 × 2)	+	0

Total preliminaries for valuation number 2 = £56,320.00

Examples 10.4–10.6 show that the methods of valuing preliminaries are diverse and for this reason the method adopted for any particular project is a matter for agreement between the respective quantity surveyors in charge of the preparation of the monthly interim valuations. To avoid later disagreements on this issue, it is always better for the client's quantity surveyor to seek the contractor's quantity surveyor's agreement on the proposed method of valuing preliminaries at either tender evaluation stage or before contract.

Unfixed material/components

These are made up of materials and components stored on or off site. The value of materials stored on site can be assessed by asking the clerk of works and the site manager to produce an agreed list for the interim valuation. On the other hand, the contractor's quantity surveyor can prepare the list of materials/components on site for checking by his or her counterpart. Failing that, both quantity surveyors can prepare and agree the list of materials/components on site during the interim valuation meeting. In addition, the client's quantity surveyor should check that none of the materials listed has been brought to the site prematurely and is not in excess of requirement and, moreover, that all materials are properly stored and protected.

All materials and components which are intended for the works but are stored off site require the architect's consent before their valuation and inclusion in the interim valuation. Where that consent has been given, the client's quantity surveyor should inspect the materials to ensure that the following conditions prevail:

- The materials/components are intended for incorporation into the works and are in conformity with the contract specification.
- The materials/components have been set aside and are clearly marked for the works.
- Ownership rights have been established in a written contract for their supply so that, when paid for, the ownership of the materials/components transfers to the client.
- The materials/components are properly and adequately insured.
- The materials/components are adequately stored and protected.

The client's quantity surveyor should also remember that the materials and components stored off site have yet to arrive on site and, therefore, the cost of transportation and returning empty carriages should be deducted from their total valuation.

Provisional quantities

Where provisionally measured items contained in the bills of quantities have been executed and measured, they should be included in the interim valuation.

Dayworks

Dayworks normally arise as a result of an architect's instruction varying sections of the works. If these variations cannot be properly valued by measurement and pricing, they should be valued on a daywork basis. It follows, therefore, that some of the architect's instructions would have been executed on a daywork basis and, in that case, the client's quantity surveyor should check whether the following are in place before their valuation and inclusion in the interim valuation:

- The architect has issued a written instruction in respect of the work considered under dayworks.
- All daywork sheets have been signed by the clerk of works or the architect.
- Daywork is the correct method of valuing that work.
- The prime cost rates are correct and the percentage additions claimed agree with that contained in the bills of quantities.
- There are no inconsistencies regarding quantities of labour, materials and plant.

Where the client's quantity surveyor is satisfied that all the above is in order, the value of the daywork is included in the interim valuation.

10.6.2 Nominated sub-contractors' application for payment

All nominated sub-contractors should be informed in advance of the interim valuation dates to enable them to submit their payment applications on time. When nominated sub-contractors submit their applications, the client's quantity surveyor checks each application against the amount of work executed and materials/components delivered to site and the appropriate figure added for each sub-contractor. It is a good practice for the client's quantity surveyor to notify each sub-contractor of the amount included in the interim valuation in their favour.

10.6.3 Nominated suppliers' application for payment

Like nominated sub-contractors, all nominated suppliers should be given advanced information of the interim valuation dates to enable them to make a timely application for payment. On receipt of any application, the client's quantity surveyor inspects all the materials/components delivered to site and checks them against the nominated suppliers' applications. The quantity surveyor must also be satisfied that the materials are of the right quality and quantity and are properly stored and protected before including them in the interim valuation.

10.6.4 Statutory authorities' application for payment

Application for payment received from statutory authorities for work executed or for materials/components supplied in connection with the work are checked by the client's quantity surveyor. These works and materials/components are included in the interim valuation if the client's quantity surveyor is satisfied with their authenticity.

10.7 FLUCTUATIONS

Construction project contracts are normally referred to as either firm or fluctuating price. Where a project has been placed on a fluctuating price contract, this means that, in case of increased costs of materials and wages during the contract period, contractors and sub-contractors receive financial reimbursement. Conversely, in the case of a fall in price of same, the client should benefit from decreased cost. It therefore follows that, in a fluctuating price contract, the amount of fluctuation must be calculated and added to or subtracted from monthly interim valuations. Traditional methods and formula methods of calculating price fluctuations are available for reimbursing the contractor or the client for increases and decreases in price at the production phase of a construction project.

10.7.1 Traditional method

The traditional method of calculating fluctuations is intended to calculate the actual increases or decreases to the contract sum as a result of changes in cost of construction resources. Under this method in the UK, the contractor's tender is assumed to have been based on rates of wages and emoluments promulgated by the National Joint Council for the Construction Industry at the date of tender. It is further assumed that the contractor has included in the tender all future increases or decreases which have been announced before the date of tender. Moreover, the contractor is required to submit with the tender a basic price schedule of materials upon which the tender price is based and, hence, on which price fluctuations are claimed. Generally, the traditional approach to fluctuations seeks to calculate the cost increase or decrease on construction resources incurred by the contractor during the execution of the works, and where fluctuations do occur during the progress of the works, they are dealt with in the manner given below.

Labour

Labour considered under this method covers the following categories of contractor's employees:

- Site operatives.
- Site staff (e.g. site engineers, surveyors and so on who are classified as craftspersons in the calculations).
- Operatives off site and producing goods/components for use on the works.

Where increases or decreases in the basic rates of labour have occurred, these are evaluated by acquiring a copy of the weekly time sheets from the contractor (verified by the clerk of works) and calculating the man-hours at the increased or decreased rate.

This procedure will also apply to fluctuations to such items as National Insurance contributions, tool money and holiday credits. The net amounts of any increases or decreases to be added to or deducted from the contract sum are included in the interim valuation.

Materials

The basis of material cost adjustment is the basic price list prepared by the contractor (or the client's quantity surveyor for the contractor to price at tender stage) and submitted with the tender. For this reason, the contractor's claim for reimbursement as a result of material price increases must meet the following conditions:

- The material must be on the basic price list.
- The price fluctuation must be due to a market price change or tax change.
- The acceptable increased price is the one relevant to the dates of material delivery to site.

For the foregoing reasons, the client's quantity surveyor does not entertain material price fluctuations arising from contractor's supply of different brands of materials, purchasing materials in small quantities or changing suppliers without tangible reasons. Apart from these exceptions, the contractor can claim for the increases or decreases in the basic cost of materials which can be evaluated from supply invoices for such materials and calculating the difference in cost between them and those contained in the basic price list. Materials considered under this method also include electricity and fuel, and, like labour, the net amounts of any increases or decreases to be added to or deducted from the contract sums are added to or deducted from the interim valuation.

Advantages and disadvantages

The advantages and disadvantages of the traditional fluctuations method may be summarised as follows:

Advantages

- Reimbursement under the traditional fluctuations method is based on actual cost incurred by the contractor.
- It is simple in approach and its use is widely understood.
- It lends itself to detailed audit and, hence, the contractor must provide proof of claim (e.g. invoices and time sheets).

Disadvantages

- Its calculation tends to be tedious, time consuming and costly to adopt.
- It takes time for contractors to get invoices and so forth together; this delays recovery and thus affects the contractor's cash flow.
- The contractor suffers a shortfall in recovery as cost items such as preliminaries, overheads and plant charges are not considered for cost adjustment.

10.7.2 Formula method

The formula method is not intended to calculate the actual increases or decreases in the cost of construction resources. Rather than the concern for actual differences in cost, the formula method makes theoretical assessment of the amount of fluctuations inherent in the value of work executed at the interim valuation period by using formulae based on statistical indices of prices and value of work done.

It is a simple and quick means of establishing reimbursement due to the contractor or client as a result of increases or decreases in the cost of construction resources. This method is based on output as opposed to input and it also lends itself to computer application.

In the UK, the adoption of the formula method causes work items in the bills of quantities to be allocated to work categories or identifiable sections to correlate the work categories contained in the bulletin published monthly by the PSA (Property Services Agency of the Department of Employment). This bulletin gives index numbers for each of the work categories (currently there are forty-nine categories) and the index numbers are intended to be used for adjusting sums included in the interim valuations. In order to apply the formula method in calculating fluctuations, the client's quantity surveyor adopts the following measures:

- Division of the bills of quantities by annotation into work categories corresponding to that contained in the PSA monthly bulletin.
- Identification and definition of a base month (i.e. base index) which is the date specified in the contract documentation.
- Preparation of interim valuations on the same day/date of each month and dividing the valuation into work categories.
- Valuation of work executed is adjusted using the formula:

$$\text{Valuation of additional work} = \text{Net value of current work executed} \times \left(\frac{\text{Current index} - \text{Base index}}{\text{Base index}}\right)$$

A worked calculation is shown in Example 10.7.

Example 10.7 Formula method of adjustment

Base index for work category 6 (say) 40

Current index for work category 6 (say) 45

Net value of work executed during the current month = £160.00

Calculation: $\dfrac{£160 \times 5}{40} = £20.00$

- Variations executed during the current month and priced at bills of quantities rates are adjusted in a similar manner. However, where variations are priced at the current rates, no adjustment is required.
- Where there is an overrun on the contract programme, the current month and all future interim valuations will be subject to adjustment using index numbers applicable at the date which the practical completion was agreed to fall.
- The value of unfixed materials and components are not considered from the adjustment.
- Provisional sum for work paid at current prices (such as daywork, work executed by statutory authorities and claims for losses and expenses) are not considered for adjustment.
- Values other than measured work such as preliminaries and attendances on nominated sub-contractors (known as the balance of adjustable works), but not allocated to work categories, are treated separately for the interim valuation.

The advantages and disadvantages of the formula method may be summarised as follows:

Advantages

- It effects savings in time and cost in its application as it is less onerous than the traditional method.
- It lends itself to computer techniques and/or application.
- Contractors can tender with more confidence as they do not need to satisfy any of the conditions associated with the traditional method.
- It causes less delay in reimbursement and, hence, provides contractors with a healthy cash flow.
- It is more equitable as calculations are based on the value of work executed and, hence, all resources are covered in the adjustment.
- Contractor benefits from employing his or her expertise in the purchase of materials and components.
- Contractor stands to gain from over recovery if index increases more than the cost of the actual material or component used on the project.

Disadvantages

- The idea that recovery of costs is not related to costs directly incurred has been a source of some concern to contractors.
- The contract sum is adjusted whether or not the contractor has incurred increased costs, and this sometimes lead to over-recovery; it is therefore rather generous and to the detriment of the client.
- Under-recovery results when the actual cost of material or components is more than the increase contained in the relevant index.
- There exist non-adjustable elements which are not considered for cost adjustment under the formula rules.

Over/under-recovery associated with the formula method

Unlike the traditional method, the formula method applies to all aspects of construction work except unfixed materials. To many, this is a fairer and quicker cost reimbursement approach. However, it has been criticised by some as a method which leads to over- and under-recovery. These variations in recovery associated with the formula method may be attributed to the disparity of data used in compiling the indices and data on actual site operations. The factors that may lead to these disparities can be summarised as follows:

- Ratio of plant, labour and materials included in the index and on site diverge and, hence, there is a difference in the amount recovered by use of the formula adjustment and the actual production costs incurred by the contractor.
- Proportion of materials used in compiling an index and quantity used on site differs, as there are occasions when the material used on the project forms only a small part of the constituent of an index or, as is usual, is not even represented.
- Variations in basic cost of resources used for the index and that on the construction project due to regional factors thus, making the contractor pay

more or less for resources than the national average on which the index is based.

- Variations in cost increases of resources used for compiling an index and those on site due to regional factors, again making the contractor pay more or less than the index stipulates.

Example 10.8 Interim valuation

UPPER DAGNALL STREET DEVELOPMENT PROJECT

INTERIM VALUATION STATEMENT FOR ARCHITECT'S CERTIFICATE

Quantity Surveyors:
Beke and Osei Partnership
28 Try Road
London, N17 4BY

Architect:
Upmarket Designers
177 Bakers Street
London, SW12 7BE

Contractor:
Fast Build Limited
Fast Build House
London, SW16 4QB

Valuation No: 4
Date:

	£
1. Value of builder's work executed (including preliminaries)	325,021.31
2. Add/omit net value of variations executed	11,044.56
	£336,065.87
3. Value of unfixed materials	75,000.00
4. Value of nominated sub-contractor's work	140,000.00
5. Value of nominated supplier's goods	37,000.00
	588,065.87
6. *Less:* Value of work or material not being in accordance with the contract	Nil
	588,065.87
7. *Less:* Retention (5%)	29,403.29
	588,662.58
8. Fluctuations	6,337.42
9. Direct loss and/or expense	Nil
	565,000.00
10. *Less:* Total of previous certificates (Nos 1–3 inclusive)	445,000.00
Amount recommended for Architect's Certificate	**£120,000.00**

SUMMARY

The post-contract cost control process is the responsibility of both the client's quantity surveyor, and the contractor's surveyor, but for diverse objectives. However, it is a process which is undertaken to ensure that the project is running according to plan and to budget. Furthermore, the preparation of an interim valuation enables the building contractor to receive payment covering the value of work properly executed. It also enables the contractor to obtain funds needed to finance the works in progress. After preparation of an interim valuation, the contractor's surveyor compares cost with value while the client's quantity surveyor compares the value of valuations with the client's cashflow forecast. For this reason, the accuracy of these interim valuations are important to avoid over- and under-valuation of works satisfactorily completed and also to provide the contractor with the needed cash flow to progress with the works. An interim valuation statement for an architect's certificate is shown in Example 10.8.

FURTHER READING

Ahenkorah, K. 1993 Re-thinking the retention rule. *Chartered Quantity Surveyor*, November, pp. 8–9.

Hibberd, P.R. 1991 Certification. *CIOB Technical Information Service, No. 126.*

Minogue, A. 1992 How to minimise your exposure to off site goods. *Building,* January, pp. 32–3.

Nisbet, J. 1979 Post contract cost control: a sadly neglected skill. *Chartered Quantity Surveyor*, January, pp. 24–8.

Powell, J.M. 1992 Certificates and payments: The new way forward. *Constructional Law*, pp. 13–17.

Ramus, J.W. 1993 *Contract practice for quantity surveyor.* B.H. Newness.

Rouse, D.J. 1986 Interim payments. *Chartered Quantity Surveyor,* October, p. 27.

Turner, D.F. 1983 *Quantity surveying practices and administration.* George Godwin.

Willis, C.J.; Ashworth, A. 1990 *Practice and procedure for the quantity surveyor.* BSP Professional Books.

CHAPTER 11

PREPARATION OF FINAL ACCOUNT

11.1 INTRODUCTION

The final account represents the final gross amount obtained after financial adjustment in respect of variations, prime cost and provisional sums, provisional quantities, dayworks, fluctuations and contractual claims at the end of a construction contract. The production and agreement to a final account usually lead to the following:

- The client's ascertainment of the total financial commitments (i.e. the total amount payable to the contractor).
- The contractor's awareness of the total entitlement under the terms of the contract.
- The issue of the final certificate by the architect for payment which, if not challenged, becomes the final payment.
- The agreement of both the client and the contractor to adjust the contract sum.
- Awareness that the effects of extensions of time granted under the contract have been considered and settled.

The above is an indication that before the final payment is made to the contractor, the final account must be formally prepared and agreed. The responsibility for preparing the final account lies with the client's quantity surveyor. However, the contractor's quantity surveyor must also agree to it and for this reason, as sections of the final account are completed, the client's quantity surveyor forwards copies of it (plus all back-up information) to the contractor's quantity surveyor for perusal, comments and/or agreement. Where major disagreements arise, these are normally solved through a series of meetings, discussions and negotiations between the respective quantity surveyors until agreement is reached and signed.

11.2 DOCUMENTS FOR FINAL ACCOUNT

Prior to commencement of the preparation of the final account, the client's quantity surveyor needs to cross-reference all sources of information, including the following:

- Architect's instructions (including site verbal/written instructions and clerk of works' instructions).
- Copies of minutes of the project team meetings.
- Daywork sheets.
- Wages sheets and material invoices (for traditional fluctuations).
- Contract bills of quantities and the correspondence file.
- Original drawings used in the preparation of bills of quantities.
- Revised architects and engineer's drawings.
- The original taking-off dimensions for the bills of quantities.
- Records of nominated sub-contractors'/suppliers' work.

In addition to the above, another good source of construction information is the clerk of works' site diary. Being regularly on site, the clerk of works is able to record most of the construction information useful for the preparation or checking of the final account and resolution of contractual disputes (when they arise).

11.3 COMPONENTS OF FINAL ACCOUNT

The preparation of the final account falls under the following headings:

- Variations account
- Adjustment of prime cost sums
- Adjustment of provisional sums
- Adjustment of provisional quantities
- Dayworks account
- Fluctuations in costs of labour, materials and statutory contributions
- Contractual claims.

11.3.1 Variations Account

The variations account consists of measured items grouped as *omissions* and *additions* under architect's instructions. Generally, where variations in quantity and/or specification occur during the progress of the works, the financial effect of these variations are dealt with. This process involves the omission of the relevant items from contract bills of quantities and the addition of the value of new items specified.

Prior to the commencement of preparation of the variations account, the client's quantity surveyor must ensure the existence of an architect's instruction which authorises the varied works. However, where the variations have arisen as a result of errors in the contract bills of quantities, these can be measured and adjusted without an architect's instruction. Furthermore, the client's quantity surveyor must inform the contractor's quantity surveyor of the intention to prepare the variations account and give him or her the opportunity to be present. Where site measurements are needed, it is good practice for both quantity surveyors to measure together in order to avoid later disagreements over the measurements. Furthermore, to simplify measurement and valuation of variations to the contract, the following procedures are adopted:

- Special double-column bill paper is issued for entering and pricing omitted (omissions) and additional (additions) items of work (see Examples 11.1 and 11.2).
- The measurement and valuation of each varied item of work is started on a new sheet of double column bill paper.
- The architect's instruction reference number and a résumé of what the instruction is about are stated at the beginning of the measurement of each varied item of work.

Omissions

Omissions are a group of items contained in the contract documentation which are no longer needed and, hence, require total deletion or substitution. The omissions are effected by simply referring to various pages in the contract bills of quantities and omitting the relevant items completely (see Example 11.1). However, in some cases, it may be necessary to refer back to the original taking-off dimensions in order to establish the actual amount of work to be omitted. In that case, it may mean copying measured information directly from the original taking-off sheets and referencing (see Example 11.2). Particular attention should, however, be paid in order to avoid omitting more quantities than originally measured or included in the contract bills of quantities, especially where omission of deductions are concerned. Moreover, the items omitted from the contract bills of quantities are normally priced at the rate contained in the contract bills of quantities.

Additions

Additions are either substituted items (quantity and/or specification) or simply extra items required for the construction project. The items of additions will often have to be measured from revised drawings. In the absence of revised drawings, these additional works are measured on site. Where applicable, items added to the contract are priced at the rates contained in the contract bills of quantities. However, if an item does not entirely correspond with that contained in the contract bills of quantities, then its price is assessed pro-rata to the item closest to the description. Where there are no like items in the contract bills of quantities, then a rate for the item must be calculated from first principles of pricing. In the latter two cases, the rate inserted is known as a *star rate*. Star rates are finally confirmed upon agreement with the contractor's quantity surveyor, who normally provides copies of invoices for materials and records of labour involved in any new item to assist the calculation and agreement of a fair rate (see Examples 11.1 and 11.2 for variations account).

On completion of the measurement and pricing of all authorised variations, a summary of all the architect's instructions is prepared and transferred to the summary of the final account.

11.3.2 Adjustments of prime cost sums

Prime cost sum (PC) is a sum of money written into the contract bills of quantities for either works which are required to be carried out by a nominated sub-contractor or goods and materials required to be obtained from a nominated supplier. In the final account, the adjustment of prime cost sums involves omitting the sums contained in the contract bills of quantities and setting against them the agreed invoices or accounts for the work executed and/or materials/components supplied. In the process, the client's quantity surveyor takes into account the main contractor's profits (where applicable), cash discounts, general and special attendance and, hence, the following distinctions require consideration.

Supply-only items

This occurs where the architect wishes to reserve the right of nominating the source of supply of a particular material/component for the works and, hence, a prime cost sum is inserted in the contract bills of quantities purely to serve that purpose. As a result, a sum only is included in the contract bills of quantities for the supply of the material/components, giving the architect the advantage of choosing the particular specification or type of materials/component he or she proposes to use at a date late in the contract. A typical example of a nominated supply item is ironmongery and the supplier selected becomes known as a *nominated supplier.*

Supply and fix items

Under this arrangement, a prime cost sum is inserted in a contract bills of quantities when the architect wishes to nominate a particular firm to carry out certain specialist works on the project. In this case, during the progress of the works, the architect acquires quotations from various specialist firms and instructs the main contractor to place an order with the firm chosen. The firm selected then becomes known as a *nominated sub-contractor.*

Main contractor's profit on prime cost sums

When the contractor prices the item of profit following a prime cost sum in the contract bills of quantities, he or she normally refers the profit to the value of the sum as a percentage. This percentage is calculated on the gross amount of the prime cost sum and, subsequently in the final account, the actual invoice inclusive of any cash discount.

Price for general attendance

The general attendance is priced as a lump sum of the prime cost sum. For this reason, in normal circumstances, the attendance item would not be adjusted against the actual invoice in the final account.

Price for special attendance

Special attendance is priced according to the amount of labour and materials/ equipment involved in providing such attendance as required for the nominated sub-contractor's work. However, special attendance will only be adjusted in the final account if the amount of labour, materials/equipment involved has changed considerably as a result of an architect's instruction.

Cash discount

Cash discounts are allowed upon nominated sub-contractors'/suppliers' invoices by way of remuneration for the main contractor's time spent in placing orders, general liaison involved in the sub-contracting, paying sub-contractors and so on. However, a cash discount is deductible only when the main contractor pays the sub-contractors/suppliers within the stated period. Most standard forms of contract allow the main contractor $2\frac{1}{2}$ per cent cash discount of nominated sub-contractors' work. This discount is deductible only when the contractor pays the sub-contractors within the stated period after issue of an architect's certificate (usually 17 or 24 days). In the case of nominated suppliers, the main contractor is allowed a 5 per cent cash discount which is deductible when he or she pays the nominated suppliers within thirty days from the end of the month during which the materials/ components were delivered to site.

During the preparation of the final account, the client's quantity surveyor pays particular regard to nominated sub-contractors/suppliers final invoices and makes the appropriate adjustment before inclusion in the final account. The types of invoices of concern are:

- Submission of nett invoices.
- Submission of invoices without the correct cash discounts.

Examples 11.3–11.6 show how the client's quantity surveyor adjusts the incorrect invoices during the preparation of the final account.

11.3.3 Adjustment of provisional sums

A provisional sum is a sum of money written into contract bills of quantities for work or for costs which cannot be defined or detailed due to uncompleted design. Therefore, financial adjustment of provisional sums covers a multitude of items which, owing to lack of adequate information, could not be measured in detail at the contract bills of quantities preparation stage. Like the variations account, the

Example 11.1 Variations account

			Description	Qty	Unit	Rate	Omissions £	p	Additions £	p
			UPPER DAGNALL STREET DEVELOPMENT PROJECT							
			FINAL ACCOUNT **VARIATIONS ACCOUNT**							
			ARCHITECT'S INSTRUCTION No. 17							
			Substitution of 200 mm thick blockwork for 100 mm thick							
			OMIT							
			Bills of quantities items:							
			4/10/J–K				3,245	36	–	–
			5/10/A–B				595	98	–	–
			ADD							
			Blockwork: Thermalite High strength 7; in mortar mix 1 : 1 : 6; as Clause F250C.							
	61.00 ×2.75		Walls: 200 mm thick.	289	m²	21.75	–	–	6,285	75
	44.00 ×2.75	167.75 14.00 288.75								
2/½/22⁄7/5.00	× 2.75/	43.20	Ditto; curved to 5.00 m radius.	43	m²	24.37	–	–	1,047	91
						£	3,844	34	7,333	66

Example 11.2 Variations account

Timesing	Dimensions	Squaring	Description	Qty	Unit	Rate	Omissions £	p	Additions £	p
			UPPER DAGNALL STREET DEVELOPMENT PROJECT							
			FINAL ACCOUNT VARIATIONS ACCOUNT							
			ARCHITECT'S INSTRUCTION No. 17							
			Substitution of 200 mm thick blockwork for 100 mm thick							
			OMIT							
			Blockwork: Thermalite							
			High strength 7; in mortar mix 1 : 2 : 6; as Clause F250C. (Original dimensions, columns 175–176 inclusive).							
			Walls: 100 mm thick							
			as bills of quantities Item 4/10/J.	154	m^2	11.24	1,730	96	-	-
	12.00									
	×2.75	33.00								
	44.00									
	×2.75	121.00								
		154.00								
2/$\frac{1}{2}$/$\frac{22}{7}$/	5.00		Ditto; curved to 5.00 m radius							
	× 2.75/	43.20	as bills of quantities Item 5/10/A.	43	m^2	13.86	595	98	-	-
			ADD							
			Blockwork: Thermalite as before described.							
	6.10									
	×2.75	16.78	Walls: 200 mm thick.	138	m^2	21.75	-	-	3,001	50
	44.00									
	×2.75	121.00								
		137.78								
2/$\frac{1}{2}$/$\frac{22}{7}$/	5.00		Ditto; curved to 5.00 m radius.	43	m^2	24.37	-	-	1,047	91
	× 2.75/	43.20								
							£2,326	94	4,049	41

Example 11.3 Invoices and cash discount

Nett invoice

The following are typical examples of invoices received from a nominated sub-contractor and nominated supplier:

INVOICE

Sinkit Deep Limited
Specialist Piling Contractors
2B High Street
Anytown, BB9 5SU

Date: ....

Supplying materials, equipment and installing 120 number 450 mm diameter cast in place concrete piles for the sum of £115,214.00 nett (exclusive of VAT).

INVOICE

Akan Architectural Ironmongers Trading Limited
74 Hill Drive
Kent, AA7 45S

Date: ....

Supplying sundry ironmongery as attached schedule for the sum of £14,750.00 nett (exclusive of VAT).

The above invoices require the following financial adjustments prior to their incorporation into the final account.

1. *Financial adjustment for cash discounts – nominated sub-contractor's account*

	£
Sinkit Deep Limited	
(Piling)	
Nett invoice dated .... for cast in place pile	115,214.00
Add: Builder's cash discount, 1/39 (to represent a $2\frac{1}{2}$% discount)	2,954.21
	£118,168.21

2. *Financial adjustment for cash discounts – nominated supplier's account*

Nett invoice	£
Akan Trading Limited	
(Ironmongery)	
Nett invoice dated .... for the supply of ironmongery	14,750.00
Add: Builder's cash discount 1/19 (to represent a 5% discount)	776.32
	£ 15,526.32

Example 11.4 Invoices and cash discount

Incorrect cash discounts
The following are typical examples of incorrect invoices received from a nominated sub-contractor and nominated supplier:

INVOICE
Sparks & Shocks Limited
Electrical Contractors
Asafo Market Road
Downtown, ST7 4JE   Date: ....

Supplying materials, equipment and carrying out electrical installation for the sum of £168,250.00 inclusive of 5% builder's discount (exclusive of VAT).

INVOICE
Wilmot & Sons
Sanitary Fittings Specialist
Park Street
Nima, OA19 9DE   Date: ....

Supplying sundry sanitary fittings as attached schedule for the sum of £9,045.00 inclusive of $2\frac{1}{2}$% builder's discount (exclusive of VAT).

The above invoices require the following financial adjustments prior to their incorporation into the final account.

1. *Financial adjustment for cash discounts – nominated sub-contractor's account*

	£
Sparks & Shocks Limited (Electrical Installation)	
Invoice dated .... for electrical installation	168,250.00
Less: Incorrect builder's cash discount, 5%	8,412.50
	£159,837.50
Add: Correct builder's cash discount 1/39	4,098.40
	£163,935.90

2. *Financial adjustment for cash discounts – nominated supplier's account*

	£
Wilmot & Sons (Sanitary Fittings)	
Invoice dated .... for the supply of Sanitary Fittings	9,045.00
Less: Incorrect builder's cash discount, $2\frac{1}{2}$%	226.13
	8,818.87
Add: Correct builder's cash discount 1/19	464.15
	£ 9,283.02

Example 11.5 Adjustment of prime cost sums

Description	Omissions £	p	Additions £	p
UPPER DAGNALL STREET DEVELOPMENT PROJECT				
FINAL ACCOUNT				
ADJUSTMENT OF PRIME COST SUMS				
ELECTRICAL INSTALLATION				
OMIT				
Bills of quantities items: 3/4/E-H.	175,000	00		
ADD				
Messrs Sparks & Shocks Limited final agreed invoice dated .... for Electrical Installation.	-	-	163,935	50
Add for profit 5%.	-	-	8,196	78
Add for general attendance (as bills of quantities item 3/4/J).	-	-	3,000	00
Add for special attendance (as bills of quantities item 3/4/K)	-	-	1,500	00
(*Note*: Similar treatment will be given to Messrs Sinkit Deep Limited account for Piling Installation in the Final Account)				
£	**175,000**	**00**	**176,632**	**28**

Example 11.6 Adjustment of prime cost sums

	Omissions £	p	Additions £	p
UPPER DAGNALL STREET DEVELOPMENT PROJECT				
FINAL ACCOUNT				
ADJUSTMENT OF PRIME COST SUMS				
OMIT				
Bills of quantities item 3/6/A–B.	10,000	00	-	-
ADD				
Messrs Wilmot & Sons final agreed invoice dated .... the supply of sanitary fittings.	-	-	9,283	02
Add for profit 5%.	-	-	464	15
(*Note:* Similar treatment will be given to Akan Trading Limited account for the supply of Ironmongery Final Account).				
£	**10,000**	**00**	**9,747**	**17**

adjustment of provisional sums involves omitting sums contained in the contract bills of quantities and adding back the detailed designed work. The items of additions which correspond with any items contained in the contract bills of quantities are priced at the bill rates. However, if an item does not entirely correspond with that contained in the contract bills of quantities, then its price is assessed pro-rata the item closest to the description. But where there are no similar items in the contract bills of quantities, then a rate for the item must be calculated from first principles of pricing. In the latter two cases, the rate inserted is known as a *star rate*. Star rates are finally confirmed upon agreement with the contractor's quantity surveyor who normally provides copies of invoices for materials and records of labour involved in any new work item to assist arriving at a fair rate (see Example 11.7 for adjustment of provisional sums).

11.3.4 Adjustment of provisional quantities

The adjustment of provisional quantities comprises the omission of items in contract bills of quantities qualified as *provisional* and the adding back of the actual remeasured items. The remeasured items are normally priced at the rates contained in the contract bills of quantities (see Example 11.8). However, star rates are negotiated where the remeasured items differ, by reason of specification and/or quantities, from those contained in the contract bills of quantities.

11.3.5 Dayworks account

All daywork sheets submitted by the contractor are examined to determine whether the items of work they contain can be properly measured and priced. If not, they are valued as daywork. In the daywork valuation, the client's quantity surveyor should be satisfied that the labour, materials and plant recorded on the daywork sheets are reasonable for the items of work involved. In addition, he or she will also check the arithmetic and ensure that the daywork valuation neither overlaps with the measured work valuation nor does the reverse by leaving a gap in the final account. The total of the daywork sheets is transferred to the final account summary and the prime cost sum for daywork is omitted from the contract bills of quantities (see Examples 11.9 and 11.10 for a typical daywork sheet and daywork account).

11.3.6 Fluctuations

As a continuous process from the interim valuations, the client's quantity surveyor will check the details of contractors' claims for increases in cost of labour and materials before inclusion in the final account.

11.3.7 Contractual claims

Where contractual claims for additional payments under the contract conditions have been accepted, evaluated and/or agreed, they are included in the final account (contractual claims are discussed fully in Chapter 12).

Example 11.7 Adjustment of provisional sums

Dimensions		Description	Qty	Unit	Rate	Omissions £	p	Additions £	p
		UPPER DAGNALL STREET DEVELOPMENT PROJECT							
		FINAL ACCOUNT							
		ADJUSTMENT OF PROVISIONAL SUMS							
		OMIT							
		Work to internal courtyard as bills of quantities item 3/1/F.				14,500	00	-	-
		ADD							
45.00 ×20.00 × 0.15	135.00	Excavate to reduce levels; maximum depth not exceeding 1.00 m.	135	m^3	3.50	-	-	472	50
		&							
		Disposal; excavated material; off site.	135	m^3	6.25	-	-	843	75
20.00 ×20.00 × 0.15	60.00	Pea shingle fillings; average thickness over 0.25 m.	60	m^3	18.50	-	-	1,110	00
25.00 ×20.00 × 0.15	75.00	Topsoil; filling to make up levels; not exceeding 250 mm average thick; obtained off site.	75	m^3	25.20	-	-	1,890	00
25.00 ×20.00	500.00	Surfaces of topsoil; grading to falls.	500	m^2	0.95	-	-	475	00
		&							
		Cultivating; surface of ground; 150 mm deep; removing debris; weeding	500	m^2	0.35	-	-	175	00
		&							
		Imported turf; BS 3969; maintaining for 12 months after laying.	500	m^2	7.15	-	-	3,575	00
					£	**14,500**	**00**	**8,541**	**25**

Example 11.8 Adjustment of provisional quantities

			Description				Omissions £	p	Additions £	p
			UPPER DAGNALL STREET DEVELOPMENT PROJECT							
			ADJUSTMENT OF PROVISIONAL QUANTITIES							
			PRECAST CONCRETE LINTELS							
			OMIT							
			Bills of quantities items: 6/7/J-L.				490	14	-	-
			ADD							
			Precast concrete lintels; BS 5328; as Clause F140.							
15/	1	15	Lintels; 100 × 150 × 1200 mm long; rectangular Section 15.	15	Nr	12.00	-	-	180	00
26/	1	26	Ditto; 100 × 150 × 1500 mm, do.	26	Nr	13.50	-	-	351	50
						£	**621**	**14**	**531**	**00**

Example 11.9 Daywork sheet

DAYWORK/RECORD SHEET

Fast Build Limited
Upmarket Street
London, SW16 2DF
Tel: 0181-000 1234

PROJECT: UPPER DAGNALL STREET DEVELOPMENT PROJECT

Sheet No: 1 Job No: 374 W/E ___ AI No: 7

Description of Work Executed:

Cutting down facing brickwork, lowering gas and electrical boxes and building at new level.

Trade	Name	M	T	W	T	F	S	S	Total	Rate	£	p	£	p
Bricklayer	D. Smith	8							8	6.50	52	00		
Bricklayer	C. James	8							8	6.50	52	00		
										Total	104	00		
										+ % 200	208	00	312	00

MATERIALS

Quantity	Description	Price	£	p	Qty	Description	Price	£	p	£	p
	Hand made Sandringham						B/F	20	00		
	Red Facing Bricks 50 No	0.40 each	20	00							
						Total		20	00		
		C/F	20	00		+ % 20		40	00	24	00

PLANT

Quantity	Description	Rate	£	p	£	p
1	Clipper Brick Cutter (for one day)	6.75	6	75		
		Total	6	75		
		+ % 15	1	01	7	76
				Total £	**343**	**76**

Client's Representative P. BLACK
Contractor's Site Representative S. ANIM
Date ______

Comments

Example 11.10 Daywork account

Description	Omissions £	p	Additions £	p
UPPER DAGNALL STREET DEVELOPMENT PROJECT				
FINAL ACCOUNT **DAYWORKS** **OMIT**				
Dayworks as bills of quantities items 3/8/A-F complete.	14,950	00	-	-
ADD				
Fast Build checked and approved daywork sheets as follows: *Sheet No.* *Value £* 1 343.76 2 1,375.50 3 180.14 4 615.18 5 331.17 6 285.02 7 525.77 8 1,326.04 9 188.17 10 246.28 11 802.49 12 1,951.42 13 441.65 14 1,116.23 15 658.59 16 1,394.05	-	-	11,781	46
£	**14,950**	**50**	**11,781**	**46**

Example 11.11 Final account summary

	Omissions £	Additions £
UPPER DAGNALL STREET DEVELOPMENT PROJECT		
FINAL ACCOUNT		
SUMMARY		
CONTINGENCIES	20,000.00	-
VARIATIONS ACCOUNT	151,137.14	167,430.76
ADJUSTMENT OF PRIME COST SUMS	763,823.00	811,214.73
ADJUSTMENT OF PROVISIONAL SUMS	176,227.00	106,956.17
ADJUSTMENT OF PROVISIONAL QUANTITIES	13,927.10	15,564.18
DAYWORKS ACCOUNT	14,950.00	11,781.46
FLUCTUATIONS	-	37,250.75
£	**1,140,064.24**	**1,150,198.05**

Example 11.12 Statement of final account

UPPER DAGNALL STREET DEVELOPMENT PROJECT

STATEMENT OF FINAL ACCOUNT

	£
CONTRACT SUM	2,729,243.14
Less: Omissions	1,140,064.24
	1,589,178.90
Add: Additions	1,112,947.30
	2,702,126.20
Add/~~*Deduct*~~: Fluctuations	37,250.75
	2,739,376.95
Add: Direct loss and/or expense	75,118.05
	2,814,495.00
Deduct: Liquidated and ascertained damages	Nil
TOTAL FINAL ACCOUNT	2,814,495.00
Less: Previous payments (as Certificate Numbers 1-18 inclusive)	2,802,495.00
BALANCE DUE TO CONTRACTOR	£ **12,000.00**

We Fast Build Limited hereby consents that, the above total in the sum of **£ 2,814,495.00** (exclusive of Value Added Tax) is the agreed final account for this contract and that the nett amount of the sum of **£ 12,000.00** (exclusive of Value Added Tax) represents the final balance due to ourselves in full and final settlement of our account.

For and on behalf of:	*For and on behalf of:*
BROADWAY DEVELOPERS LIMITED	**FAST BUILD LIMITED**
Signed ______________________	Signed ______________________
Date ______________________	Date ______________________

SUMMARY

In this chapter, we have discussed the preparation of the final account. The preparation and agreement of the final account concludes the contract between the client and the contractor. The final account is normally prepared by the client's quantity surveyor and this is checked by the contractor's surveyor. At times, both surveyors prepare it together and the process involves measurement of new items, remeasurement of varied items and adjustments of the various prime costs and provisional sums.

FURTHER READING

Ahenkorah, K. 1994 Reimbursement for changes in construction costs. *The Building Economist*, March, pp. 19-22.

The Aqua Group 1982 *Contract administration for architects and quantity surveyors*. Granada.

Nisbet, J. 1979 Post contract cost control: a sadly neglected skill. *Chartered Quantity Surveyor*, January, pp. 24-8.

Ramus, J.W. 1993 *Contract practice for quantity surveyors*. B.H. Newness.

Turner, D.F. 1983 *Quantity surveying practices and administration*. George Godwin.

Wainwright, W.H.; Wood A.A.B. 1979 *Variation and final account procedures*. Hutchinson.

Willis, C.J.; Ashworth, A. 1990 *Practice and procedure for the quantity surveyor*. BSP Professional Books.

CHAPTER 12

PROBLEMS IN CONSTRUCTION

12.1 INTRODUCTION

The complexity of today's construction projects brings together many skills and conditions which are always variable. In the atmosphere of complexity and uncertainty, some construction projects do not run as smoothly as planned and, hence, exposes the client and the building contractor to many problems. While some of the construction problems can be solved without the expenditure of time and money, the solution of others requires the expenditure of time and/or money and, moreover, the involvement of the courts. The process of solving construction problem without recourse to court action is known as a contractual claim. However, this process is useful where both parties are trading and can afford the extra expense, and therefore it is not useful where one party to a contract is insolvent.

12.2 CONTRACTUAL CLAIMS

Contractual claims are assertions of the right to remedy or relief by a party under the express or implied provisions of the particular contract. More simply, a claim is a request for compensation for damages incurred by any party to a contract. It can be made within the contract or at common law and the purpose of a contractual claim is to place the claimant in the position he or she would have been in had the breach or cause of the claim not occurred.

Most standard forms of contract recognise the complexity and uncertainty in construction projects and, hence, contain clauses which permit the contractor to claim for direct loss and/or expense suffered due to occurrence of certain events under a contract. These clauses, therefore, provide the means whereby any dispute involving the particular event can be resolved without resorting to litigation or involving the courts. Therefore, during the production phases, if the contractor recognises that any of the events has occurred and, as a result, he or she has suffered direct loss and/or expense, a detailed record of such events is to be forwarded to the architect immediately as a basis of claim for reimbursement. The notification of any claim for reimbursement must be forwarded as early as possible to ensure that the architect is aware of the additional expenditure on the project and also to provide him or her with an opportunity to check or identify the event causing it. In addition, the contractor must also be aware of his or her contractual responsibility to mitigate any effects of the problems encountered rather than leave it uncontrolled, thinking that, at all costs reimbursement will result at the end of the day.

12.2.1 Prolongation and acceleration claims

A prolongation claim covers a contractor's request for financial reimbursement when production time is extended by the client's action or request. Acceleration claim, on the other hand, refers to a contractor's request for reimbursement of costs incurred when he or she consents to double productive efforts in order to reduce the timescale of programmed activities and/or prevent delay. It follows, therefore, that if a client's action causes production delay, he or she can either accept a revised completion date and pay for a prolongation claim or ask the contractor to accelerate production and thereby pay for the acceleration cost. The decision of construction clients faced with this dilemma will be influenced by the need to complete the project on time. In projects where time is of the essence, acceleration cost is likely to be less than the cost of prolongation due to the period that can be reduced by acceleration. In addition, when production is in delay, a decision to accelerate will enable the project to meet its original completion date. In other times, too, from the client's points of view, it may be advantageous to opt for the acceleration of the project's production for the following reasons:

- Where, for commercial reasons, it is of vital importance to achieve completion by the original or an earlier date.
- Where the benefits to be accrued from acceleration outweigh the costs.
- Where the actual loss to the employer for late completion exceeds the liquidated damages which may be recovered from the contractor.

12.2.2 Preparation of contractual financial claims

The responsibility for the preparation of the contractual financial claim rests with the contractor's quantity surveyor. However, what triggers this activity is the occurrence of an event which causes the contractor to suffer direct loss and/or expense. When this situation has arisen, the contractor's quantity surveyor must act as follows:

- Identify the terms and conditions that are expressed or implied under the contract apply.
- Show that the client has breached the identified terms and conditions.
- Demonstrate the provisions in the contract document that entitle the contractor to reimbursement.
- Prove how the contractor has suffered direct loss and/or expense and demonstrate by calculation the constitution of the direct loss and/or expense.

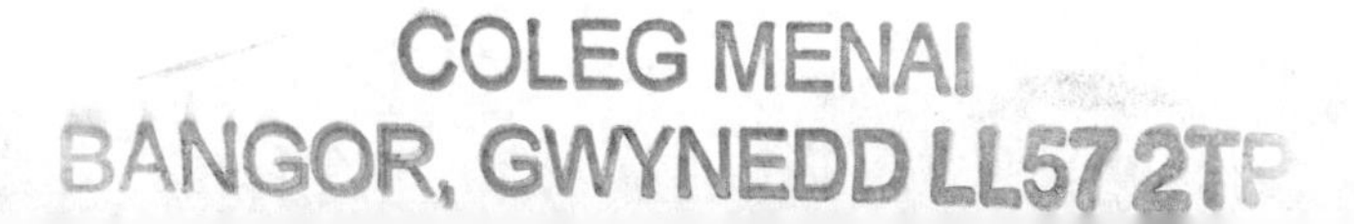

The success of a contractor's claim for reimbursement for the costs of delay and disruption depends to a large extent upon accurate and sufficient information in support of the claim. Indeed, standard forms of contract usually impose this requirement as an obligation upon the contractor. However, such information will not be readily available unless adequate records are kept as production progresses. Therefore, the maintenance of adequate records is a major prerequisite to compilation of a successful financial claim, and in order to carry out the above functions effectively the contractor's quantity surveyor may need to examine and/or refer to the following documentation (in no particular order of importance):

- All correspondence including minutes of project team meetings.
- Architect's instructions and directives issued in writing.
- Site instructions delivered by the architect or clerk of works.
- Contract/working drawings.
- Updated and new drawings and other drawn information.
- Site diary and daily weather reports, showing overall weather pattern for the duration of project.
- Daily labour allocation sheets showing daily labour utilisation on all sections of the works.
- Contractor's tender build-up showing breakdown of unit rates into labour, plant, materials, profit and so on.
- Authorised daywork records/sheets complied by the contractor during the progress of the works.
- Original construction programme.
- Updated construction programme due to changes requested by the client.
- Plant and scaffold records showing hire and use of plant.
- Material schedules indicating amount and time of incorporation into the works.
- List of invoices received and payments made to sub-contractors and suppliers.

After assembly and examination of the above records, the contractor's quantity surveyor progresses the preparation of a contractor's financial claim in stages as follows:

1. *Contract particulars:* Under this heading, the contractor's quantity surveyor gives the description of the works, contract period, dates (such as that for tender, commencement, original completion, practical and actual completion), and extension of time granted.
2. *Points of claim:* The points of claim comprise a list of causes of the claim in respect of delay, disruption and extra costs upon which the contractor is seeking recompense under the express and implied terms of a particular contract. Points of claim are diverse but may include the following statements:

   (a) Issuing of variations to such an extent and at such times that they cause disruptions to the regular progress and sequence of the construction works.
   (b) Failure of the architect to approve drawings submitted by nominated specialist sub-contractors and/or suppliers on time to enable the works to make a steady progress as planned.

(c) Issuing of variations/instructions and/or construction information too late to allow their easy incorporation within the programmed sequence of the works.
(d) Disruption suffered by a contractor as a result of operatives/artisans employed directly by the client.

12.2.3 Detailed calculation of claim

The contractor's quantity surveyor, as a rule, calculates the contractor's financial claim for direct loss and/or expense under a number of heads of claim. The heads of claim vary and are normally identified from the list of points of claim. The detailed calculation of a typical contractor's financial claim may be carried out under the following headings:

Prolongation

Prolongation costs arise as a result of the original contract period being extended by either an issue of extensive variations or by a client's action which disrupts the regular progress of the works. Cost items considered under financial claim for prolongation are:

1. *Preliminaries:* Under this cost heading, such items as the cost of time-related preliminaries (e.g. site staff salaries, attendant labour, plant and standing scaffolding, small tools, site huts, watching and lighting, telephone and electricity costs) increase as the contract period prolongs. The contractor's quantity surveyor will consider and include the amount of actual direct loss and/or expense on preliminaries suffered by the contractor resulting from the extended contract programme.
2. *Changes in period of working:* If, as a result of the prolongation, the contract is extended into the winter months or a rainy season, contractors generally incur additional cost on items such as precautions required for winter working, reduced working hours, productivity stoppages through bad weather, drying out and other associated costs. The record of these costs and other details normally form part and parcel of a contractor's financial claim and, hence, it is considered accordingly and included in the financial claim.
3. *Additional financing costs:* Additional financing costs cover a financial claim for interest on outstanding monies (i.e. a retention fund) which a contractor has been prevented from utilising due to the prolongation. However, this cost comes under the definition of direct loss and/or expense only when a contractor had submitted the right notices. Where this requirement has been complied with, the contractor's quantity surveyor will include the cost under this heading of the financial claim.
4. *Increased costs:* Where tenders obtained for the construction project were on a firm price basis, contractors do suffer direct loss and/or expense if the cost of construction resources (e.g. labour, materials) increase as the production period

prolongs. If that should be the case, the contractor's quantity surveyor will consider and calculate the actual increases on resources due to inflationary pressures and include them under this heading of the financial claim.

5. *Attendance:* Where the prolongation of a contract period leads to extension of attendance (general or special) that a contractor normally provides on nominated sub-contractors under a contract, the actual costs of the extra period of attendance is considered, calculated and included in the contractor's claim.
6. *Loss of profit:* A contractor's financial claim for loss of profit represents a demand for the profit which a contractor would reasonably have expected to earn from other work had the prolongation not occurred. Generally, a contractor's financial claim for recovery of profit is related to the required percentage return on turnover. But as this percentage had already been expressed in a contractor's tender submission, the contractor's quantity surveyor applies the same percentage in the calculation of the direct loss and/or expense under this heading of claim. However, this claim must be supported by a demonstration that, as a direct result of the disruption and/or prolongation, the contractor has been prevented from earning profit elsewhere in the normal course of business. The contractor must be able to prove, for example, that it had not been possible to respond positively to invitations to bid for available work because his or her resources were retained on the delayed project.
7. *Head office overheads:* Financial claim under this heading covers cost associated with head office personnel and equipment (e.g. head office buildings, office machines, office running, maintenance and staffing costs) required for the contractor's business. Generally, these items are a fixed cost a contractor incurs whether or not he or she trades, and every project a contractor undertakes is expected to contribute towards the maintenance of these overheads. For this reason, the head office overheads are usually calculated annually for the coming year and expressed as a percentage on the anticipated turnover for that trading year. The percentage is then used as the markup on all tenders submitted in that year. It follows, therefore, that if a project is delayed beyond the scheduled completion period, a contractor is unable to achieve the anticipated level of turnover and, hence, this leads to a shortfall of income for financing the head office overheads. In claiming for the additional overhead cost resulting from a prolongation, the contractor's quantity surveyor needs to provide proof of the contractor's annual turnover, expected return on turnover and the percentage of this return required to cover the head office overheads.

Acceleration

Under construction contracts, acceleration means expediting production pace to meet the original or revised contract completion date. If the project is running late, the client may require a contractor to take steps to complete it within the original programme. Alternatively, the contractor may unilaterally decide to take measures to avoid delays and payment of liquidated damages. If the client wishes to achieve completion earlier than anticipated, he or she may instruct the contractor to accelerate. Therefore, a contractor will normally act to make up the production time when instructed by the client to do so. Nevertheless, if the contractor is in culpable delay, he or she will have the right to choose between paying liquidated damages or increasing expenditure by accelerating. Usually, the increased productivity obtained from the acceleration is achieved by either the introduction of additional site resources (manpower and plant), or prolonged working hours achieved by overtime and/or shift working arrangements. The above moves aimed at saving production time are expensive to implement and therefore the cost of any acceleration would be assessed by the contractor's quantity surveyor for inclusion in the contractor's financial claim.

To facilitate calculation and evaluation of an acceleration cost if a client instructs a contractor to accelerate site production, the agreement of the contractor must be secured. An agreement to accelerate is by way of either a variation to the terms of the original contract or a separate agreement. It is therefore essential to state clearly what is required, and at what cost, to minimise the risk of future problems or disputes. For this reason, the cost of this financial claim is usually agreed by both parties before the commencement of the accelerated production. Moreover, negotiations for this package is extremely difficult to conclude as it normally entails many assumptions and guesswork on the probable cost of accelerating. However, when an agreement has been reached the contractor receives periodic payments for his or her cost through an architect's certificate.

The alternative approach to the calculation of this financial claim is the identification of the heads of cost and agreement on a method of calculating the extra cost of the acceleration. For example, it may be agreed that the contractor submits daily labour returns for assessment of the amount of labour on site, overtime hours worked and so on; and that periodic cost calculations and payment should flow from the information/records provided.

Disruption

A financial claim for disruption is a direct loss and/or expense claim which arises as a result of client's variations on the project having extensive disruptive effect on site production. When, for example, an architect's instruction creates an uneconomic/out-of-sequence working and/or loss of productivity of labour and plant, the contractor incurs an additional cost in the production. Moreover, in order to maintain the required level of output or rate of progress in a project facing disruptive effects of variations, a contractor may need to reprogramme the works, reallocate resources, reorder materials, return unwanted materials and so forth many times within the project's life. In addition, under such conditions, plant and labour cannot work to full capacity; hence, output targets as programmed may not be achieved and the material wastage factor may increase. The preparation of a financial claim for direct loss and/or expense arising from disruption is an extremely difficult exercise and the contractor's quantity surveyor relies, to a great extent, on the contractor's site records, the programme, etc., in order to prepare and submit a successful financial claim under this heading. However, a client's quantity surveyor's clear knowledge of the site operations normally facilitates the settlement of this claim.

12.2.4 Ascertainment of contractor's financial claims

The responsibility for ascertainment of the amount of a contractor's financial claim for direct loss and/or expense rests with the architect unless the client's quantity surveyor is specifically instructed in writing to do so on the architect's behalf. The ascertainment involves a check to verify that the contractor's financial claim is valid and that the integral parts of it are admissible and have been correctly valued. During the process of the ascertainment, a client's quantity surveyor goes through each heading of the financial claim thoroughly to ascertain the validity of each section. Where the supporting information is inadequate, he or she may request, from the contractor, the additional information required to substantiate the claim. At times, too, the information required – such as salaries of site staff, directors' fees and so forth – may be acquired by the inspection of relevant documents in a contractor's head office. Generally, the ascertainment of most items will not present a client's quantity surveyor with any difficulties. For example, the cost of time-related items such as site accommodation, plant, electricity, telephones, etc., are normally easy to evaluate from a contractor's records and invoices. However, items such as head office overheads, disruption, loss of profit and increased costs are seldom straightforward. These will require a careful consideration and a contractor's cooperation in the supply of the required information in order to conclude the ascertainment and/or negotiation satisfactorily.

12.3 INSOLVENCY

Insolvency generally refers to the state of being unable to meet one's debts when they are due and creditors are pressing for payment. Therefore, it arises whenever a person, partnership or company is unable, because of a lack of assets or other resources, to pay debts as and when they fall due for payment. Insolvency of the contractor or of the client is a tragedy that can befall any construction project, but the structure of the construction industry makes it particulary prone to that eventuality. When a company falls into that unfortunate situation, a person known as a *liquidator* is appointed to carry out the process of winding it up. The liquidator's duty is to realise the property of the insolvent company, pay its debts and distribute the rest (if any) among its shareholders/employees.

12.3.1 Insolvency of the contractor

The peaks and troughs in construction orders resulting from the state of the national economy exposes contractors to the risk of insolvency. Generally, contractors do well in buoyant economies due to optimism in the money market, housing and commercial property boom but fare badly in recession. When a national economy is in recession, contractors' order books become empty and the shortage of orders leads to keen competition in tenders with a low or nil profit markup. The result of this practice is high claims for direct loss and/or expense which is intended to recoup some or all of the financial cost/benefit not included in the tender. In that case, the success or failure of such excessive claims determine the financial position of the contractor; which also means that in that type of economic climate, many contractors continue to trade when they may be totally insolvent.

There are other important factors which make the contractors susceptible to insolvency, and these include the following:

1. *Under-valuation:* Contractors suffer cash flow difficulties owing to the failure of the client's quantity surveyor to value accurately for payment of works satisfactorily completed. The persistence of this practice can strain the contractor's ability to provide adequate funds for both the project and his or her business, and can also drive the contractor into insolvency.
2. *Retention fund:* High retention funds and failure of clients to release them when due can frustrate the contractor's performance in business.
3. *High interest rates:* In firm price contracts, contractors stand to suffer from any upward movement of interest rates which they failed to foresee and, hence, failed to allow adequately for in their tenders.
4. *High cost of resources:* In firm price contracts, contractors are deemed to have allowed all future cost increases of resources resulting from inflationary pressures. Hence, they stand to suffer from any inadequate allowance for these increases in their tender.
5. *Mis-management:* Contractors suffer losses due to bad business management. Their fixed overheads may be high and, therefore, need a substantial turnover to be sustained. In addition, their site performance may be poorly planned. This may result in loss of production time or defective workmanship which is costly to remedy.
6. *Insolvency of client:* Failure of clients to pay for work properly executed by reason of insolvency is one of the contractor's worst fears. The amount of money owed may be substantial and this can drive the contractor and his or her sub-contractors out of business.

When the contractor becomes insolvent, the client's project comes to an immediate halt in mid-course, followed by a whole range of disputes and problems. To protect the client's interest during the production stage, it is advisable that the client's professional advisers remain alert and look for signs which indicate that the contractor may be heading for an insolvency situation. The following factors therefore require attention:

- A sudden slow-down of production and/or reduction in value of materials on site without any tangible reason.
- Premature removal of plant and/or tools from the site.

- An undue delayed payment to nominated sub-contractors/suppliers and statutory authorities and, finally, failure to honour certificates.
- Withdrawal of sub-contractor's labour and frequent changes of sources of material supply.
- Changes in site staff and supervisory operatives.
- High proportion of sub-standard/defective workmanship.
- Unusual visits to site by contractor's senior management.

When any of the above signs are noticed, an appropriate investigation must be carried out to establish the facts of the matter and, where necessary, measures taken immediately to safeguard the client's interest. The measures appropriate in the circumstances include the necessary preliminary plans to have the site, works and materials protected, ensuring accuracy of interim valuations, and asking the contractor to provide proof of payment to nominated sub-contractors/suppliers. Detailed plans should also be prepared for the completion of the works.

12.3.2 Action upon insolvency

Until the contractor is officially or legally insolvent, the client must be informed of the contractor's situation by his or her professional advisers and advised to continue the performance of all contractual obligations under the contract. The client must therefore, honour all interim certificates that become due to the contractor and must not enter the site to take control of events. However, if the insolvency becomes a reality, the client should be advised of the automatic determination of the contractor's employment and the measures being adopted to complete the project with a minimum of delay. The client should be given information on the stage the contract has reached and the financial position, site safety and security arrangements, insurance, additional cost, estimated date for completion and notice to bondsman (where the contractor is bonded). In addition, it must be remembered that, due to his or her insolvency, the determination of the contractor's employment has automatically taken place but the contract with the client is still alive and, hence, the financial implications of the insolvency need sorting out. Therefore, the client's quantity surveyor will need to carry out the following activities:

- Check the last interim valuation and make further technical assessment, to ensure that:
  - Any over-valuation is corrected in the current interim valuation (if any).
  - An adequate allowance is made in the current interim valuation (if any) to cover any remedial works.
- Prepare, in conjunction with the architect, a schedule of defective work which should be copied to the client, liquidator and bondsman (where applicable).
- Prepare, in conjunction with the architect, an inventory of all temporary buildings, plant, tools, equipment, materials and so on left on site with an indication of ownership.

- Prepare a schedule of cost items which should be charged to the contractor. This schedule may include the following items:
  - Rectification of defective work.
  - Replacement of defective, deleterious or deteriorated materials.
  - Removal of disused plant from site.
  - Site safety and security between determination and recommencement of production.
  - Temporary works to protect half completed works.
  - Hire charge of essential plant and equipment left on site.
  - Completion of uncompleted work which requires completion before recommencement of production.
  - Additional professional fees and expenses for the completion contract.
  - Tidying up site for recommencement of production.
- Prepare a schedule of sums of money attributable to the contractor. This list will include retention funds, value of work satisfactorily completed, work undertaken since last interim valuation, possible under-valuation and value of any direct loss and/or expense claim.

Getting the project completed

There are four options open to the client in getting the project completed. These are as follows:

- Allowing the insolvent contractor to complete the project through a novation agreement.
- Re-tendering for the completion of the construction project.
- Allowing the insolvent contractor to proceed until a novation agreement is reached, or retender.
- Allowing the insolvent contractor, after consideration and agreement, to complete the works.

Novation

Novation is a contractual device whereby, with a client's express approval, a new contractor steps into the shoes of an insolvent contractor. In effect, the new contractor purchases the interests in the project and, for a consideration, assumes all the rights, benefits and liabilities of the contract. The completion of the project by novation arrangement enables the client to continue the project without any additional costs. However, this is possible only where the novation agreement clearly makes the new contractor liable for the following:

- General making good of defects (patent and latent, if applicable) in the works.
- Carrying out all emergency work.
- Rectification of any damage (deterioration or act of vandalism) resulting from the determination.

- Responsibility for the time lost between the determination and recommencement of production.
- Arrangement for the assignment of nominated sub-contractor's/suppliers' contracts from the insolvent contractor to the new contractor.

Retendering

This is a completion contract arrangement whereby the remainder of the client's project is put out to tender. In the interest of speed, it is always considered prudent to approach the second lowest tenderer on the original tender list and see if he or she is prepared to take over the project at the original tender rates offered. Where that organisation is unwilling to take over the project on the original tender rates, negotiation of revised rates should be considered. The alternative is to obtain new tenders from interested contractors, although this approach takes time to organise as the client's quantity surveyor will need to prepare, expeditiously, project completion documentation. There will also be a need for the architect to amend the original contract drawings to suit the new contract before invitation of tenders.

On receipt of tenders, the client's quantity surveyor carries out the usual evaluation and make a recommendation to the client for his or her approval. When a new contractor is appointed and production has recommenced on site, both the architect and the client's quantity surveyor perform their normal contractual functions (i.e. contract and financial administration) until completion. However, on completion, the client's quantity surveyor will have two final accounts to prepare - the normal final account and a notional final account.

Notional final account

The notional final account is a representation of expenditure that should have been incurred by the client had the determination of the insolvent contractor's employment not occurred. For this reason, it is prepared on the basis of the original contractor having completed the whole works. This is then compared with the actual total cost including client's direct loss and/or expense (if any) resulting from the determination. The difference of the two final accounts represents a debt due either to or from the client. In the process of the preparation of this final account, all works satisfactorily executed are valued using cost information or rates contained in the original contract bills of quantities. The same treatment is given to valuation of variations and daywork items and the notional final account is presented in a format similar to that shown in Example 12.1.

This example shows that, owing to the insolvency of the contractor, the client spent £3,072,325 on the development project instead of £2,783,828. The client would therefore endeavour to recover the additional expense of £288,497 that had been incurred from the insolvent contractor, and if there is a bond holder, he or she will be liable for 10 per cent of £2,728,043, which is equal to £272,804. If that should be the case, the client will seek to recover £272,804 from the bond holder and will be an unsecured creditor for the remaining £15,693, which should be notified to the liquidator.

Example 12.1 Notional final account

1. Final account for insolvent contractor

	£
Contract sum	2,728,043
Add net adjustments:	
(i) Architect's instructions nos 1–60 executed by insolvent contractor	17,645
(ii) Architect's instructions nos 61–100 executed by new contractor but priced at insolvent contractor's rates	35,000
(iii) Prime cost and provisional sums	3,140
Total final account	**£2,783,828**

2. Amounts paid for project's completion

	£	£
(a) Final account for completion contract		
New contract sum	1,400,000	
Add net adjustment:		
(i) Architect's instructions nos 61–100 executed by new contractor and priced at his or her rates	42,375	
(ii) Prime cost and provisional sums	3,750	1,466,125
(b) Gross valuation for interim payments to insolvent contractor (as architect's certificate nos 1–10)		1,500,000
(c) Extra expenses incurred by the client		
(i) Nominated sub-contractors paid direct	96,200	
(ii) Nominated suppliers paid direct	39,500	
(iii) Watching and protecting site	5,500	
(iv) Additional insurance	2,500	
(v) Additional professional fees	27,500	
	£ 171,200	
Less: Plant sold on behalf of insolvent contractor	20,000	151,200
(d) Damages claimed by client		50,000
		£3,147,325
Less: Retention 5% of £1,500,000		75,000
		£3,072,325
Less: Insolvent contractor's final account		2,783,828
Client's total net expense		**£ 288,497**

Contractor carrying on until novation agreement or retender

In projects where time is of the essence, this arrangement can often prove useful if the client can be assured that:

- The insolvent contractor will not remove materials, items of plant and equipment from the site.
- The insolvent contractor has materials and manpower capacity to carry on.
- Nominated sub-contractors/suppliers will be paid for their work.
- Materials and work will be protected from acts of vandalism and the ravages of weather.
- The insolvent contractor will tidy up the site prior to the takeover by the new contractor.
- Liability for making good defects rests with new contractor.

It must be mentioned, however, that this temporary arrangement will be under the control of an insolvency practitioner and it is only a time saver if proper arrangements can be made to put it into practice.

Insolvent contractor finishing the contract

This procedure is adopted when by agreement between the client, the insolvent contractor and the liquidator, the contractor's employment is reinstated. This arrangement is adopted where the works are nearly completed and substantial sums of money are due to the contractor. However, this approach can create problems for the client if, subsequently, there are several items of defective work to be made good.

Generally

When a project ends up in insolvency of the contractor, the client's professional advisers should select a method which will enable recommencement of site production with a minimum of delay. The reason for this is that, the greater the period that elapses from the date of determination of the insolvent contractor's employment and recommencement of site production, the greater the likely additional costs due to deterioration, theft, acts of vandalism, etc. Also, the project completion method selected is influenced by the production stage of the project prior to the determination, and the following may be given as a guide to the appropriate action necessary to reflect the project's stage of production:

- Where work has been started but no substantial amount of work is complete, consideration should be given to the acceptance of the second lowest tenderer's bid after negotiation.
- Where work has been started and the project is half complete, consideration should be given to negotiating with the second lowest tenderer, use of novation contract approach or retender.
- Where work is substantially complete, consideration should be given to the selection of a new contractor on negotiated rates or daywork basis.

12.3.3 Insolvency of the client

Like contractors, clients sometimes become insolvent and the signs of a client's approaching insolvency include:

- Late honouring of architect's certificates.
- Unnecessary arguments over payments to the contractor.
- Spurious set-offs without identification or quantification.
- Requests to defer payments.
- Failure to pay the contractor.

To safeguard his or her interest, if any of the above should happen, the contractor should ask for confirmation that the retention fund has been held in a separate account as stipulated in the contract documentation and/or as requested. When the threat of the client's insolvency becomes a reality, the contractor must decide quickly whether or not to end employment with the client. If the contractor decides to determine his or her employment, the client should be given written notice to that effect. Thereafter, the contractor has the right to stop work, remove temporary buildings, plant, equipment, materials, etc., form site. He or she must advise all sub-contractors to do likewise and provide them with the facilities to do so. Thereafter, the contractor's quantity surveyor will normally establish the total amounts owing to the contractor under the contract and under the determination as a result of the client's insolvency. The amount due from the client is the sum total for value of work executed, direct loss and/or expense, materials on site, cost of removal of site huts, plant and so on and also damages for loss suffered due to the determination of his or her employment.

SUMMARY

In this chapter, contractual claims and their evaluation have been discussed, together with the problem of insolvency. Whilst a contractual claim can be presented under several headings, the primary classification is whether the claim is the result of prolongation or acceleration. The former claim arises as a result of the extension of production time while the latter is the result of a client (after, for example, a disruptive action) wanting his or her project on or before the agreed completion time. In either case, the client pays for his or her action in the form of acceleration cost or direct loss and/or expense the contractor incurs as a result of the prolongation of project's production time.

Another problem the construction industry experiences periodically is that of insolvency of one of the parties to a construction contract. Generally, insolvency of the building contractor delays the production of a construction project and also exposes the client to increased cost. Insolvency of the client, on the other hand, leads to a total halt of the project with unpaid bills to the building contractor and sub-contractors down the contracting chain.

FURTHER READING

Douglas, R.W.R.; Robertson, G.F. 1978 '*Claim*' is not a four letter word. *Building Technology & Management*. April, pp. 8–11.

Farrow, T. 1991 Acceleration; the agreement. *Chartered Quantity Surveyor*, September, pp. 24–5.

Goodacre, P. 1979 Delays in construction; ascertaining the cost. *CIOB* .

Hawkins, R.A.P. 1993 Insolvency under JCT contracts. *CIOB Construction Papers, No. 24.*

Mann, L.C. 1982 Insolvency; how to get the work finished. *Construction Surveyor*, July, pp. 18–21.

RICS 1987 Contractor's direct loss and/or expense. *Practice Pamphlet No. 7.*

RICS 1984 Insolvencies. *RICS Quantity Surveyors Practice Pamphlet No. 5.*

Robinson, T.H. 1977 *Establishing the validity of contractual claims. CIOB*

Wreghitt, J.D. 1989 Claims for prolongation and disruption. *The Surveying Technician*, October/November, pp. 8–9.

CHAPTER 13

RESOLUTION OF CONSTRUCTION DISPUTES

13.1 INTRODUCTION

Construction projects are generally complex and for this reason, delays and disputes are always present. Although the client has a desire to acquire the right building, at the right time and at the right price, he or she is always exposed to possible delays and/or additional costs for which there may be no compensation.

13.2 OBJECTIVES OF PARTIES

In a construction contract, the objectives of the client are to acquire the building at the right time and at an economic price. The contractor, on the other hand, aspires to complete the building in accordance with the contract requirements at minimum cost with a view to profit maximisation. While the objectives of both parties to contract are recognised in construction projects, the complexity of a construction project brings together a wide range of individuals with varying practical and professional skills. Hence, the success of all construction projects depends on cooperative effort, efficiency and dedication of all individuals involved in its production and management.

Generally, all the participants in a construction project aim to complete it within time, cost and to specification, but this does not exclude contrasting perceptions of the project's requirements. Furthermore, there may be conflicting objectives among project participants which can also result in non-performance. This disparity in the appreciation of project objectives and in construction skill and/or experience is exacerbated by the traditional separation of design from construction, inadequate construction information on which management decisions are taken and the nature of the competitive bidding market. For this reason, claims, counter-claims and disputes in construction are inevitable and are generally the consequences of the failure of one or more project participants to fulfil their contractual obligations. While it can be said that most of the disputes which arise in construction contracts are of a minor nature and are settled quickly and satisfactorily by the parties concerned, there are some matters of dispute that are not easily resolved and, subsequently, become the subject of claim by one party against the other.

13.3 CONSTRUCTION DISPUTES GENERALLY

Over the years, in order to increase the success rate at which construction projects can be managed and produced, construction professionals have concerned themselves with the promotion of production efficiency, high level of confidence and team spirit among project participants. In the process, professional advisers make every effort to examine all contract documentation as thoroughly as possible to discover and eradicate any aspects which they consider, on application, could possibly cause a dispute.

However, notwithstanding these measures, it must be recognised that construction projects are intrinsically complicated, hazardous and at the mercy of the unpredictable – such as weather – and hence it is perhaps not surprising that something often goes wrong. For this reason, no construction project is free from problems, and when problems are not immediately solved as they arise, they can become major issues which eventually end up in court or before an arbitrator for resolution. When a construction dispute reaches that stage, its outcome becomes extremely costly for all concerned.

When the client's project becomes involved in a protracted dispute, the consequence is that it will no longer meet the original goal and expectation. In a protracted dispute, the client stands to suffer from the payment of high legal fees and delayed completion, resulting in late occupation which affects sale and letting and incurs a loss of revenue. There will also be higher production costs due to inflation and interest on borrowed funds. The contractor, on the other hand, will not be able to achieve the profit maximisation objective as he or she will suffer financial loss from unpaid work and claims, and payment of legal fees. As a result, millions of pounds per annum are dissipated in the form of contractual disputes of one sort or another in the construction industry. In most cases, however, the cost of pursuing an action very often grow to the extent that they exceed the amount in dispute. There is a need, therefore, to carry out an analysis of the possible financial outcome of an action balanced against the direct and indirect costs of pursuing that action. Only then will disputants possess the information they need to take a decision on whether a settlement should be attempted by negotiation (including conciliation) or by an arbitration award/ court judgment. Such an analysis will certainly lead to prudent decisions which may save disputants, and the construction industry as a whole, money which would otherwise have been spent on the direct and indirect cost of resolving construction disputes.

13.4 CAUSES OF DISPUTES

Disputes may arise on a construction project for a number of reasons. Some well-known ones include:

- Shortcomings, omissions and errors in contract documentation giving rise to ambiguities in contract requirements.
- Delays in the supply of general construction information.
- Late issue of instruction varying some sections of the works.
- Increase in scope of work (changes, extras and errors) without proper consideration for extension of production time.
- Untimely issue of variation instructions, which disrupts the contractor's progress and programme of works.
- Failure of contractor to construct the works diligently and to programme.
- Poor workmanship and failure to use specified materials, skilled operatives and recognised methods.
- Failure to inspect works in progress regularly and condemning only when works are completed.
- Inaccurate valuation of variations and works in progress.
- Acceleration to complete within original programme without proper agreement over the payment.
- Late or non-payment for works satisfactorily completed when payment is due.

It can be deduced from the above list that the main areas of construction disputes revolve around time and cost of overruns, quality of workmanship, payment, contract documentation, construction information and site supervision. In addition to the above classification, it can be said that, owing to their diverse status, the viewpoints of the project participants towards the disputed areas will always vary; Table 13.1 delineates some of the viewpoints of the project participants.

Table 13.1 Project participants viewpoints of disputed areas

Dispute areas	Viewpoint of project participants			
	Client	Contractor	Sub-contractor	Consultants
Time for completion	Late completion and payment for extended contract period expected	Insufficient; speed up production to avoid payment of penalty for non-completion	Under pressure to speed up production to programme	Sufficient; invocation of penalty clause should contractor default
Quality workmanship	Good quality product to specification assured	Good quality workmanship ensures future commissions	Production to specified quality	Inspection; non-payment for work below specified quality level

Table 13.1 *Continued overleaf*

Table 13.1 *Continued.*

Dispute areas	Viewpoint of project participants			
	Client	Contractor	Sub-contractor	Consultants
Payments	Uncertainty of final cost of product due to variations and claims	Prompt; should reflect production cost and aid maintenance of good cash flow	Prompt; should reflect production cost and aid maintenance of good clash flow	Avoid over-payment, keep cost of variations and claims within client's budget
Contract documentation	Should be reliable if all contentious points covered	Inadequate; ambiguous with hidden extra cost items	Only part disclosed; inadequate	Adequate; clear enough for the execution of the works
Construction information	Assured that it is adequate, clear and timely issued	Late issue by client; inadequate; causes delay and disruption	Late inadequate, disrupts production programme	Contractor possess enough information to carry on
Site supervision	Interest being protected by constant quality inspection on site	Effected to ensure good and satisfactory workmanship	Conflicting; should come from single source	Periodic site inspection effected to ensure good and satisfactory workmanship

13.4.1 Time for completion

Normally, project completion time is specified by the client, but when it is made an object of competition it is specified by the contractor. In either case, it is expected that the contractor will endeavour to complete the whole of the works within the agreed or specified time. There are occasions, however, when contractors fail to complete projects on time, and this denies clients' early disposal or use of the facility. Late completion of construction projects can lead to claims and counter-claims; and can also damage the business relationship of parties to a contract.

Typically when projects suffer delayed completion, a defaulting contractor may attempt to avoid payment of a penalty for non-completion by attributing the overrun on the contract programme to the following number of failings by the client and the client's professional advisers.

- Failure of client to give possession of site timely as agreed or specified in the contract documentation.
- Late issue by the architect of construction information with which to progress the works.

- Untimely issues of instructions which introduce new additional works and/or variations to the construction project.
- Disruption of production and construction programme by the issue of numerous variations by the architect.
- Disruption of production and construction programme by artisans employed directly by the client.
- Inequitable extension of time award by the architect for the project's completion.

In the case of an architect's failure to grant an extension of time when a delay has been caused by the client, a contractor may argue defensively that time is at large which, in simple terms, means that the completion date is no longer applicable and therefore the contractor's contractual obligation is reduced to completion of the project within a reasonable time. But what constitutes a reasonable time is subject to interpretation by a court of law. Furthermore, in addition to putting forward the above defence, the contractor may claim to have suffered direct loss and/or expense and, accordingly, claim financial reimbursement.

13.4.2 Quality of workmanship

As a rule, in construction contracts, the contractor covenants to undertake and complete construction works to a standard of quality specified at a price the client expects to pay. Therefore, a client will only be satisfied if the contractor is able to provide a finished product of good quality on time. For this reason, it is incumbent upon a contractor to execute the works diligently and display good workmanship at all times. However, there are occasions when contractors achieve speed of construction at the expense of quality. On such occasions, the quantity of output per period is regarded as more important than its quality aspects and, when the sub-standard work is challenged, in defence a contractor may put forward the following counter-accusations:

- The standard of quality required lacks clarity in the contract documentation.
- Inadequate communication at site level on issues concerning quality standards.
- The design details do not promote buildability upon which quality depends.
- Standard of quality as envisaged by the client is not included in the contract documentation.
- The construction project never received regular quality inspections during its production.

Generally, when the quality of the works or a section of it is not up to the specified standard, the contractor receives no payment for it. In such situations, the architect instructs the client's quantity surveyor to omit the value of the unsatisfactory work from interim valuations and this results in money withheld until the unsatisfactory work is remedied.

13.4.3 Payment

Payment procedures are often a major source of construction dispute. A contractor expects to be paid for work done and accordingly, in construction contracts, contractors are paid only for work properly executed. Depending on the size and duration of the contract, payment may be made either at the end of the contract (e.g. jobbing works) or intermittently through interim valuation/certificates (e.g. large projects).

Generally, a contractor who operates a systematic planning and cost control procedure is able to compare, at any point in time, what is happening in relation to what was budgeted to happen and can reconcile the cost against the value. In addition, a contractor's own internal valuation enables interim valuation levels to be determined, which is an important datum for planning cash flow requirements. While it is recognised that maintenance of good cash flow is the life blood of a building contractor's business, a contractor may experience one or more of the following payment problems which eventually disrupt all efforts to maintain the good cash flow vital for his or her business operations:

- Interim certificates which do not reflect the true value of work properly executed.
- Interim certificates issued at irregular periods by the architect.
- Failure on the part of the client to honour interim certificates on time.
- Late settlement of final account and contractual claims for direct loss and/or expense sustained.

The factors contributing to the above area of conflict are numerous, but include the following:

- Interim valuations prepared by a client's inexperienced or over-cautious quantity surveyor who is wary of over-paying the contractor.
- Under-measurement of works contained in the contract bills of quantities which does not truly reflect the amount of work executed by the contractor.
- Failure of the client's quantity surveyor to prepare interim valuation statement in good time to enable the architect to issue a certificate.
- Failure of the architect to process the interim certificate in good time.
- Neglect of the client's quantity surveyor and the contractor's quantity surveyor to value and agree authorised additional works executed for inclusion in interim valuations.
- Failure of client's quantity surveyor and contractor's quantity surveyor to prepare and settle final account in good time.
- Delays by the contractor in submitting contractual claims for direct loss and/or expense and/or details of it.
- Delays by the contractor in submitting details of increased costs on labour and materials.
- Poor performance by contractor due to the employment of an unplanned or unspecified method, untrained or incompetent operatives, inadequate and inaccurate data.

- Under-pricing of items contained in the contract bills of quantities as a result of contractor's error in estimating.

Normally in construction contracts, all works properly executed by a contractor are measured and valued at the end of a project. This is an indication that all under-payments, by reason of any of the points outlined above, are automatically rectified in the final measurement and valuation. Nevertheless, it must be stressed again that a healthy cash flow is vital for a contractor's business. It is, therefore, important that effort is spent in the preparation of all interim valuations in order to arrive at a figure which truly reflects the contractor's cost of production.

13.4.4 Adequate contract documentation

Contract documentation is the recording in formal documents of the content, terms and conditions of a contract. In construction projects, it embodies the entire written and drawn construction information; and contract conditions constitute the core of the agreement between client and contractor and, hence, comprise articles of agreement, conditions of contract, drawings, bills of quantities and/or specification. To safeguard the client's interest, the contract documentation should fully explain the offer and acceptance between the client and the contractor. It should also adequately describe the scope, quantity, quality and positions of the work and define the rights and obligations of the parties. A carefully drafted document promotes the smooth running and successful completion of a construction project. However, despite its important legal significance, a common source of complaint and dispute is the inability of the contract documentation to treat and describe project information requirements adequately. Generally, there seems to be a widespread misunderstanding of the contractor's information needs and, hence, contract documentation is issued in various erroneous states which includes:

- Inadequate drawn or written project information which does not convey clearly what is required and, hence, is incapable of being realised.
- Construction method, specification and quality standard which fail to reflect the skills, materials/components and plant readily available.
- Contract documentation which fails to disclose the complex nature of a construction project and obligations or restrictions imposed on contractor such as accessibility, position and use of site; limitation of working space and working hours; presence of existing services; maintenance of existing services; order or phases of executing works and handover/completion requirements.
- Discrepancy between various contract documents (e.g. between drawings and specifications; and between standard conditions and specially written statements in the preliminaries section of contract bills of quantities).
- Inappropriate insertion or alteration of clauses in standard forms of contract without due consideration of the effects on other contract clauses.

- Erroneous inclusion of specific items in contract bills of quantities aimed at opting out of the current standard method of measurement.
- Contract bills of quantities prepared in a rush and based on inadequate pre-tender information and, hence, lacking in accuracy of both description and quantities.
- Large items of work covered by prime cost sums, provisional sums or provisional quantities, which eventually create an erratic or unworkable construction programme.
- Erroneous inclusion of clauses in contract bills of quantities intended to override those contained in a standard form of contract in use.

The adequacy of contract information saves production time and facilitates the effective utilisation of a contractor's resources. Therefore, to prevent disputes arising under this heading, contract documentation should be carefully, adequately and accurately prepared and be consistent throughout. As a rule, additional information may be required at the production phase but these should supplement the existing obligations rather than impose others in addition to those contained in the contract conditions.

13.4.5 Production information

During the production of a construction project, further construction information is usually issued to clarify or supplement the information made available at the pre-contract stage. Also, the project architect may consider varying sections of work to suit the client's revised requirement or to rectify a design error. This new or revised information is normally presented in various forms (e.g. verbal/written instructions, descriptive schedules and additional revised drawings or details) to change or clarify those already supplied. However, very often conflicts arise from the issue of post-contract construction information in the form of architect's instructions as a result of the following flaws in the system:

- Instructions or construction information provided lacking preciseness in wording.
- Failure of instructions or the information they contain to indicate their purpose (i.e. whether new work is being ordered, alteration is required to work already executed or clarifying discrepancy).
- Instruction effected merely by issue of revised drawings without any explanation of the nature and extent of variations on the drawings.
- Late issue of instructions and/or information (e.g. after an item of work for which the variation is intended has been executed).

The practices outlined above do not promote smooth execution of construction projects; rather, they are the source of construction disputes as they create difficult working conditions, complicate purchasing and administration and upset financial arrangements. Timely, clear and adequately worded instructions or information have a positive effect on site operations and productivity.

13.4.6 Supervision

Close supervision of a construction project during its production phase is essential to the achievement of the specified standard relating to quality. On construction projects, conflicts do arise when contractors fail to supervise the works properly. Generally, neglect of proper site supervision leads to the following:

- Production of sub-standard work.
- Condemnation and additional remedial costs.
- Non-payment for defective work.
- Failure to properly control and coordinate sub-contractors' works.
- Maintenance of poor site records.
- Failure of operatives to execute the works diligently and safely.

A defaulting contractor usually blames others (mainly the project architect and the clerk of works) for poor supervision and, hence, sub-standard workmanship. Also, in defence to the accusations over poor supervision and its effects, the contractor may use the arguments discussed under section 13.4.2 above. He or she may also put forward the following assertions and counter-accusation:

- Competent qualified personnel are in charge of production supervision on site.
- Failure of clerk of works and/or project architect to pass comment on sub-standard work until completed.
- Clerk of works is awkward and uncooperative in matters relating to supervision and quality of the works.
- The architect and the clerk of works expect an unspecified standard of quality.
- Health and safety precautions were not specified in contract documentation.

The above is an indication of a dispute-prone industry in which claims and counter-claims are endemic. While these are recognised facts of a construction project's life, the client's professional advisers are expected to take measures to minimise disputes. Furthermore, the project team's awareness of a client's project strategy and their respective responsibilities from the onset of a construction project is a good approach towards reducing conflict. Through this awareness, they will be able to accept the responsibilities entrusted to them and, hence, communicate effectively and perform or respond properly and timely to construction project information as they become available.

13.5 RESOLUTION OF DISPUTES

As mentioned above, construction disputes are inevitable because a project participant may fail to perform, respond properly and timely, communicate or understand construction information. Construction contract documents are often long and complex and contain contract terms which provide guidance on courses of actions open to parties should they find that they have a difference of opinion or that a dispute has arisen. Therefore, in theory, the express term of the contract should provide solution to any contractual disputes that arise between the parties by reference to the contract terms appropriate to the circumstances. In practice, however, some difficulties do arise in the definition of the express terms and their relation to the legalities of the event or events causing the dispute; and this is likely to result in the need for a third party (court, arbitrator or mediator) to assist in the resolution of the dispute. In spite of this, when one party feels that the contractual obligations or expectations have not been met and seeks financial and/or time compensation, resolution is sought by one or more of the widely recognised methods discussed below. A classification of contractual claims and disputes is presented in Figure 13.1.

13.5.1 Construction litigation

Litigation is a dispute resolution method which involves the use of the courts of law, a third party who is trained and qualified in law, and a judge appointed by the courts. This method of construction dispute resolution can be lengthy, extremely expensive and, hence, only suitable for resolving complex disputes in which large sums of money or clarification of a vital legal point in construction contract is at stake. Moreover, as the case should be prepared properly for trial, it is inevitable that a significant period of time will elapse between the commencement of proceedings and the trial. However, generally the speed of hearing, in most cases, depends upon the following:

- Expeditious preparation of the case by parties or litigants.
- Availability of competent legal advisers to handle the preparation of the case and the hearing on behalf of parties.
- Availability of courts and judges to hear and try the case.

Depending upon the sum involved, in the UK, the court may decide to hear a construction dispute in either the County Court or the High Court. However, to avoid delays and the expense of a full hearing, the High Court may refer a case to the official referees court, especially if the case is primarily of a technical nature, in spite of the fact that points of law are also involved.

An official referee is a specialist circuit judge whose court is more adaptable to commercial procedures than a County Court or High Court and, hence, handles expediently most general commercial and construction disputes. The official referee's approach to dispute resolution is *interventionist*. They were the first judges to suggest an exchange of witness statements before trial and a meeting of experts to agree facts and narrowing issues in dispute to effect early settlement. Hence, when a construction dispute is referred to an official referee's court, the dispute does not normally involve a full hearing; rather, the official referee is asked to consider preliminary legal questions or points. His or her decision, after consideration of the preliminary legal points, enables the litigants

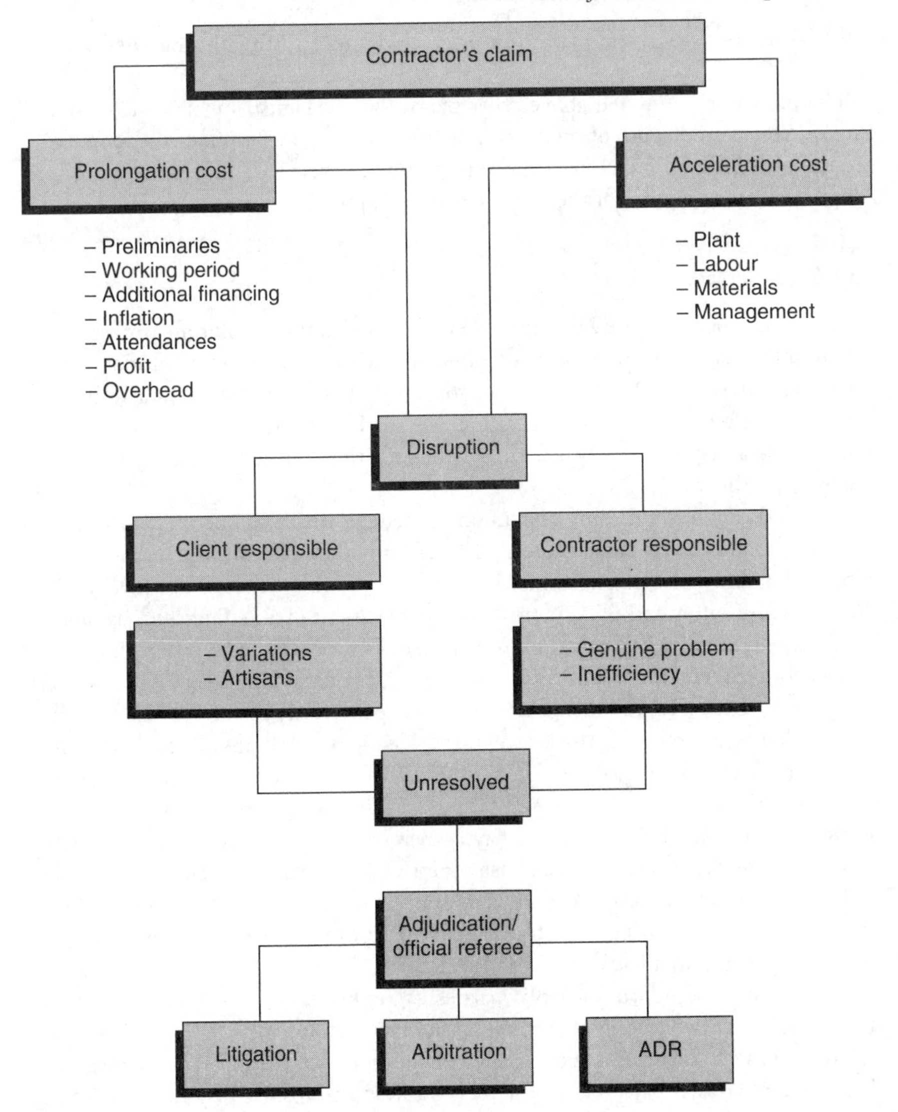

Fig. 13.1 Classification of contractual claims and disputes

to decide to either dispose of the case or proceed to a full trial. At times, too – especially where witnesses live some distance from the courts – the official referee may sit at a location convenient to the parties in dispute.

Another approach to a short, quick and inexpensive dispute resolution is the request for a summary judgment. In this approach, the plaintiff may allege that there is no arguable defence to the claim and, therefore, he or she should be entitled to judgment without a full trial of the merits of the dispute. Provided that the court is satisfied that the defendant has no defence which warrants a full trial of the issues, the plaintiff will get judgment forthwith, together with costs.

The procedures outlined above are some of the mechanisms which enable the English courts to dispose of disputes in a cost-effective way without the lengthy process of full trial. Therefore, from time to time, construction litigants in the UK adopt them in resolution of their construction disputes.

13.5.2 Adjudication

In UK construction contracts, adjudication is a dispute resolution process which authorises a third person unconnected with a construction project to decide on an issue judicially. The person entrusted with that responsibility is known as an adjudicator and, on appointment, he or she settles construction disputes that may arise between the client and the contractor during the progress of the works.

An adjudication is a dispute resolution procedure which can be used to soften a strained situation at its initial stages. The reason is that, once a dispute has hardened in the construction industry the instinct of the parties is to process it to the costly and bitter end. Therefore, early involvement of an adjudicator avoids such a situation arising. The adjudicator is usually an expert who is conversant with construction procurement and management and, hence, is able to apply this expert knowledge when needed. He or she must act expeditiously to resolve issues early when called upon to do so. Currently in the UK, some recognised standard forms of building contract already provide for the use of adjudication. These include the BPF system, GC/Works 1 and the ICE New Engineering contract and, depending on the form in use, the adjudicator may be named in the contract document, named at the onset of site production or chosen jointly by the parties when the need arises.

This system, unlike other dispute resolution systems, is not intended that there should be a formal hearing. Where necessary, parties may even put their point in dispute in writing to an adjudicator, who then has a set number of ways to respond. On receipt of an adjudicator's reply, parties can either choose to abide by his or her decision or go to arbitration at the end of site production.

This approach to construction dispute resolution is fast, cheap, takes the heat out of disputes, and enables parties to realise their position and resolve disputes early. Also, the referral of disputes to an adjudicator and the expediency of his or her reply neither prolongs disputes nor damages the cordial relationship of parties to a contract. Under this method, differences of opinion on matters such as contractor's entitlement to and length of extension of time, compliance to specification, effects of variations and their valuations, entitlement to terminate the contract and matters of valuation, can be dealt with speedily. Generally, adjudication is splendid if it is used as a means of obtaining a quick decision which is binding until overturned in arbitration. The decision may seem to be unceremonious and ready, but it will at least allow the parties to continue with the construction project in hand. They know that there will be an opportunity later, in arbitration, if they want to have the merits of their respective cases carefully and systematically weighted before a decision is made.

13.5.3 Arbitration

Arbitration is a dispute resolution method in which the solution is recommended by an impartial third person (rather than a court) chosen by the litigants themselves for that particular purpose. The exercise of freedom in the selection of a third person (known as an arbitrator) to resolve disputes does not mean that an arbitration overrides the jurisdiction of the courts. The reverse rather is true as arbitration tends to follow the rules of the High Court. For this reason and also affirmation of this fact, pleadings, reports and evidence can be expected in an arbitration hearing just as in a High Court action. However, in spite of these similarities, the two systems have some major differences which may be summarised as follows:

- In arbitration, the parties may be represented by whoever they wish as opposed to barristers and solicitors normally retained in litigations. Parties are obliged, under the principle of *discovery*, to disclose to each other the existence of all documents known to them which are relevant to the matters in dispute.
- Litigation is a procedure handled in public courts, while arbitration is handled within the construction industry. For this reason, arbitrators who are construction experts, and have also worked within the industry, travel round the country specifically to hear construction disputes and hand down decisions.

Normally, parties who intend to take their dispute to arbitration make this agreement either before a dispute arises or after it has arisen. Most standard forms for construction contracts incorporate arbitration clauses (to the effect that an arbitrator will be appointed soon after a dispute has arisen) ready for invocation when the need arises.

The conduct of arbitration

In the UK, the procedure adopted for the conduct of arbitration is set out in the Arbitration Act 1950. However, this act implies that parties have considerable freedom in deciding how they wish the arbitration to be conducted and the arbitrator is not bound to follow the formal procedures adopted by the courts. This, therefore, also gives the arbitrator the freedom to adopt procedures to suit the complexity of the dispute to be decided and the attitude of the parties in dispute. Generally, on acceptance of his or her appointment, the arbitrator arranges a preliminary meeting for the parties to discuss the following:

- Date, place and time for the hearing.
- Date for the submission of pleadings (i.e. points of claim, defence, counter-claim, etc.).
- Arrangements for discovery of documents.
- Arrangement for witness, transcripts, appearance of council, etc.

In simple terms, the procedure for a hearing consists of the claimant opening and conducting his or her case (i.e. setting out the facts and remedy sought). In defence, the respondent opens his or her case and responds to each of the claimant's allegations. This is followed by the discovery of documents, the calling of witnesses and the final address by the respondent and a reply from the claimant. After the hearing has been conducted and finished, the arbitrator must, in accordance with the law, decide on each issue as presented and publish his or her award. Arbitrators do not normally give their reasons for a particular decision unless both parties request it or the arbitration agreement requires them to do so. The arbitrator, therefore, discharges the duty for which he or she was appointed when the award is published and delivered. Table 13.2 makes a comparison between arbitration and litigation procedures of resolving construction disputes.

Resolution of construction disputes through court proceedings can be a lengthy and expensive process. Additionally, there are a number of other reasons why injured parties might be influenced to adopt arbitration rather than the courts in the resolution of construction dispute.

- Arbitration is flexible, as the parties choose the arbitrator themselves and, also, the time and place for hearing can be arranged to suit the convenience of parties.
- Arbitration proceedings are almost invariably held in private as no one except the

Table 13.2 Arbitration and litigation comparison

	Arbitration	Litigation
Duration	Expedited process from start to decision	Lengthy process from start to judgment
Cost	Saving in expense can be effected	High cost of legal representation
Privacy	Assured; held in private, public excluded	No privacy, held in public
Flexibility	Flexible; place, time and conduct to suit parties	Inflexible; place, time and conduct decided by the court
Formality/advocacy	Parties represented by anyone (matured/sane) parties may decide	Parties represented by qualified lawyers
Technical expertise	Arbitrator has expert knowledge relating to the technicalities of the dispute	High Court judge is well versed in the law but may lack expert knowledge relating to dispute
Business relationship	Business relationship may be resuscitated	Business relationship damaged by the outcome
Decision	Made private; hence not damaging	Judgment made in public and hence damaging
Settlement finality	With few exceptions, final and binding	Operates appeals procedure

parties are entitled to be present without express consent of both parties and the arbitrator.
- The arbitrator is a person versed in the technicalities of the dispute and will almost certainly inspect the subject matter of the dispute if it is appropriate to do so.
- Arbitration is less stressful than court proceedings and parties are therefore able to resolve their dispute amicably and maintain their business relationship.

While the above is a demonstration of the advantages of resolving construction disputes through arbitration proceedings, there are some disadvantages to be experienced from its adoption. These include the following:

- In litigation through the courts, there is an appeal procedure, but in arbitration the award (with the exception of a very limited right to appeal) is final and binding.
- While the arbitrator's award is final and binding on parties (subject to limited right to appeal), they are not binding on anyone else and therefore the system does not create a record of decided cases upon which anyone can rely to resolve other similar construction disputes.
- The arbitrator can be a *master of his or her procedure* and, hence, can choose to ignore the time and cost considerations and instead adopt delaying tactics in the serving of pleadings, for example.
- The arbitrator may be both unsuited and unqualified to understand many construction disputes and, hence, his or her award may have fundamental errors and may even breach the rules of *natural justice*.

13.5.4 Alternative dispute resolution

Alternative dispute resolution (ADR) is a non-confrontational technique which may resolve disputes without resorting to traditional litigation or arbitration. It is claimed to be fast, effective, inexpensive and less threatening or stressful. ADR is the Chinese traditional method of dispute resolution; however, the method has been endorsed in the USA and successfully employed in resolving disputes in Australia, New Zealand and Hong Kong. ADR offers both parties the opportunity to participate in the process and empowers them to be creative in solving their own conflicts. When a contractual dispute arises, parties are encouraged to be open minded and make every effort to resolve the dispute in the most amicable way possible. Table 13.3 compares the characteristics of an ADR approach to construction dispute resolution with that of traditional litigation/arbitration. The procedure for conducting ADR is shown in Figure 13.2.

Characteristics of ADR

ADR has proved to be a fast and economical way to settle disputes without recourse to formal proceedings. However, it relies on there being some residual trust between conflicting parties who are genuinely willing to negotiate and arrive at a mutually acceptable settlement. The peculiar characteristics of ADR are as follows:

Table 13.3 Litigation, arbitration and ADR comparison

Characteristics	Litigation	Arbitration	ADR
Place/conduct	Public court; unilateral initiation; compulsory	Private (with few exceptions); bilateral initiation; voluntary (subject to statutory provisions)	Private bilateral initiation; voluntary
Hearing	Formal; before a judge	Formal; conforming to rules or arbitration; before an arbitrator	Informal; before a third party (a neutral)
Representation	Legal; lawyers influence settlement	Legal; lawyers influence settlement	Legal; only if necessary; disputants negotiate settlement
Resolution/disposal	Imposed by a judge after adjudication; limited right of appeal	Award imposed by arbitrator; limited right of appeal	Mutually accepted agreements; option of arbitration if dissatisfied
Outcome	Unsatisfactory legal win or lose	Unsatisfactory legal win or lose	Satisfactory business relationship maintained
Time/cost	Time consuming; uneconomic	Can be time consuming and uneconomic	Fast; economic

1. *Predisposal to settle:* Before commencement of negotiation, the parties should be genuinely predisposed to effect settlement without litigation or arbitration and, hence, be prepared to compromise.
2. *Non-binding proceedings:* Proceeding are non-binding until a mutually agreed settlement is accomplished. Therefore, either party can resort to (resume) litigation or arbitration if the procedure fails.
3. *Senior management's role:* Senior management of the conflicting parties' establishments play an active role in the resolution. Even when they are represented by a lawyer, senior management make the final decision that effects the settlement.
4. *Influence of commercial interest:* Negotiations and settlement are governed by commercial interest rather than by rule of law. Therefore, ADR revolves around technical and commercial issues and seeks to maintain good business relationships.

Forms of ADR

The most common forms of ADR offer the choice of conciliation, mediation and executive tribunal or mini trial.

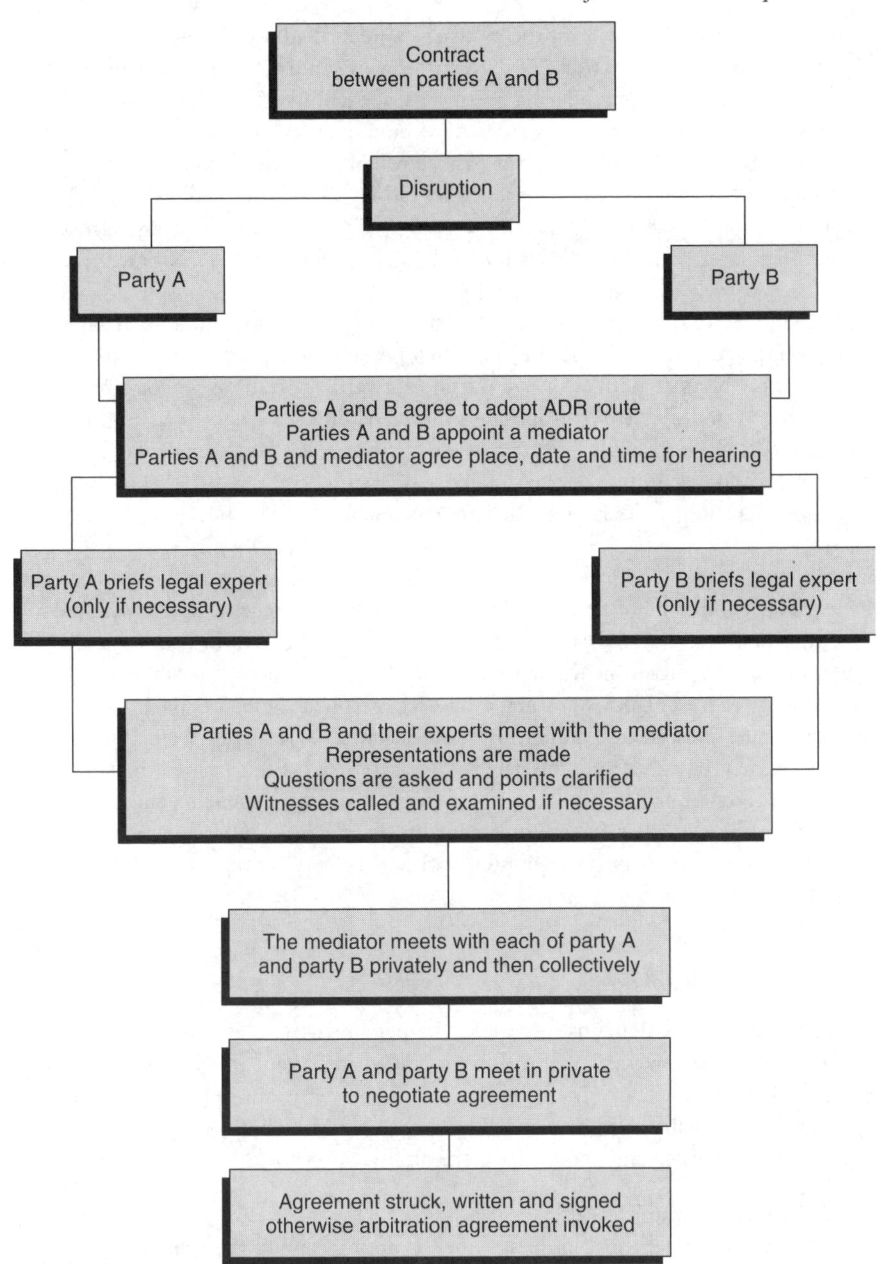

Fig. 13.2 Procedure for conducting ADR

Conciliation Conciliation is a process where a neutral adviser listens to the disputed points and endeavours to explain the views and the arguments of each party to the other. By engendering awareness of each other's problems and viewpoints, he or she encourages parties to come to an agreed solution. In this approach, the neutral adviser plays a passive role in the negotiations leading to a settlement. No recommendations are made; rather, the adviser facilitates communication and leaves the parties to settle their differences. If parties are able to reach an agreement, the adviser puts it in writing for each party's signature.

Mediation With mediation, a neutral adviser listens to representations from both conflicting parties and helps them to shape a possible settlement. The adviser, in this instance, plays an active role by putting forward suggestions, encouraging discussions round his or her proposals and persuading parties to focus on the key issues. He or she also holds private discussions with each party, explains points and more or less coaxes a mutually acceptable settlement. If successful, the adviser puts this agreement into written form for endorsement by both parties.

Executive tribunal The executive tribunal is a more formal mediation carried out by a group comprising a neutral adviser (the chair) and representatives of senior executives (who are vested with authority to settle all matters in dispute and make a binding agreement) from the management of each disputant's establishment. At the hearing, representations are made by each party to the chair. Each party may also raise questions and/or points for clarification or seek advice from the chair. Witnesses may be called but this facility is rarely used. Both parties then retire to a separate room to try to negotiate a settlement. Any agreement reached is put into writing by the chair for endorsement by the parties. Under this arrangement, more than one chair may be appointed and parties may be represented by legal experts.

Qualities of the neutral adviser

The success of ADR depends in part on the personality (i.e. assertiveness and skill) of a neutral adviser, who plays a disinterested but important role in resolving the dispute. A neutral adviser may be either a construction professional or a lawyer specialising in construction disputes. He or she should be both acceptable to and trusted by the parties and should have the ability to:

- Listen and communicate effectively.
- Develop and implement a process which is appropriate to the specific circumstances of the dispute.
- Probe, define, analyse, clarify and solve problems and the disputants' differences.
- Display interpersonal skills of setting disputants at ease, reducing defensive and aggressive behaviour and, moreover, building trust between the parties.
- Perceive or appreciate all the relevant issues in the dispute and direct the disputants' attention to the crucial points.

Functions of a neutral adviser

A neutral adviser has to employ considerable skill in achieving settlement with disputants who are normally angry, frustrated and often confused. Depending on the circumstance, the functions performed by a neutral adviser may include the following:

- Setting the scene or forum for mediation and educating parties on the process and procedures involved.
- Conducting the mediation in such a manner that neither party is seen to be losing face.
- Making sure that a party's cooperation and willingness to alter his or her position is acknowledged and appreciated by the opponent.
- Facilitating communication and helping parties to identify ways of solving their problems.
- Suggesting, clarifying, interpreting, reasoning, persuading, evaluating and informing the parties about the strengths and weaknesses of their case.
- Structuring and preparing the preliminary draft of a settlement agreement.
- Assisting in binding the parties to their agreement and monitoring its implementation.

The advantages and disadvantages of the ADR may be summarised as follows:

Advantages

- A saving in expense is effected due to a reduction in settlement time and the possible elimination of legal counsel costs.
- It is flexible in terms of absence of formalities; place, date and time can be arranged to suit the convenience of the disputants.
- The objects of the dispute and the outcome are decided by the disputing parties themselves and not by or under the influence of legal experts.
- Most construction disputes are commercially and technically oriented (rather than legal) and, in consequence, are well suited to ADR.
- Direct participation by the disputants and the understanding of the strengths and weaknesses of each other's case lends itself to a creative and amicable outcome.
- Disputants make a joint effort to find a mutually beneficial commercial solution, rather than legal solution which may be too abstract, restrictive or inappropriate to the needs of their respective businesses.
- Disputants are at liberty to find a solution which promotes continuity of on-going business relationships without loss of face.
- Conflicting parties are able to avoid unwarranted publicity and, hence, can protect their trade secrets.

Disadvantages

While the above demonstrates the strengths and benefits of ADR in resolving disputes, it is not a panacea for all construction disputes for the following reasons:

- It is only effective where disputants genuinely wish to negotiate a settlement and, hence, are prepared to compromise.

- Generally, a complex construction dispute cannot be resolved painlessly in a few days; unless, of course, one party risks a compromise which may result in financial loss for the sake of continued business relationship.
- The neutral adviser may be biased, lack understanding of the technical content of the dispute and may be without the skill and expertise required for shaping a possible settlement.
- The method is non-binding and, therefore, an unscrupulous party can enter the process of negotiation knowing that there will be no agreement but that the final payment will be delayed. He or she may also unfairly glean useful information from the discussions and disclosures made and use it later to advantage.
- ADR cannot be employed in the resolution of complex disputes which require legal opinion or in a case where public hearing or a legal precedent is required.
- A neutral adviser may experience difficulties in extracting the truth of an important event of the dispute, and this may protract the settlement to a point where litigation may be the only option.
- Disputants may be influenced by self-interest, defensiveness and legal advice on the strength of their case and, hence, be unwilling to compromise.

Generally

Although ADR has the benefit of flexibility, speed, economy and privacy, the prime factor which influences its adoption is a genuine predisposition of disputants to settle their differences without recourse to protracted confrontation. It follows, therefore, that the complexity of a dispute or the magnitude of what is at stake has no bearing on the decision to embark on the ADR route. Rather, it is the rational decision of reasonable parties to explore avenues for negotiation and thereby avoid damaging confrontation that determines its adoption. Nevertheless, this important decision is made voluntarily and mutually by the parties without a third party's influence. As it is influenced by the need to foster continued good business relationships, once this decision has been made, disputants are obliged to make every effort to agree a settlement and thereby dispose of the dispute.

SUMMARY

In this chapter we have discussed problems and disputes in the construction contracting environment. Construction projects are generally complex by reason of the number of participants associated with it and, also, the problems associated with the organisation of human resources, materials and procedures. Construction process dealing with the organisation of many human factors do not come easily to construction project leaders. This weakness brings with it failures in production plans, disputes and ways of resolving them. As mentioned above, this unfortunate situation stems from the diverse and complex nature of construction projects and, also, the absence of a tried and tested method of dealing with the complexities associated with it.

FURTHER READING

Bernstein, R.; Wood, D. 1993 Handbook of arbitration. Sweet & Maxwell.

Bevan, A. 1992 Alternative dispute resolution. Sweet & Maxwell.

Bunton, L. 1986 Contractual disputes. *Chartered Quantity Surveyor,* October, pp. 24–6.

Elliot, R.F. 1988 *Building contract litigation.* Longman.

Jawad, A. 1994 Alternative dispute resolution – mediation as a cost effective method of dispute resolution. *The Structural Engineer,* **72** (7/5), April, pp. 109–12.

Knowles, R. 1990 Adjudication is the answer. *Chartered Quantity Surveyor,* February, pp. 17 and 34

Kwakye, A.A. 1993 Alternative dispute resolution in construction. *CIOB Construction Papers No. 21.*

Mildred, R.H. 1982 Arbitration as applied to the construction industry. *Quantity Surveyor,* May, pp. 78–80.

Stephenson, D.A 1982 *Arbitration practice in construction contracts.* E&FN Spon.

Turner, D.F. 1989 *Building contract disputes their avoidance and resolution.* Longman.

CHAPTER 14

THE CHANGING CONSTRUCTION PROCUREMENT SCENE

14.1 INTRODUCTION

In the foregoing chapters, the traditional sequential method of construction procurement, its adoption and problems have been considered. Despite its imperfections, which have received several criticisms from various quarters of the construction industry, the system will probably still be in use for many years. However, as everything in nature undergoes a continuous change – and the construction industry is no exception – at some stage, procedures within its framework are bound to change. Even now some changes are evident, but the factors influencing this change are very complex. Also, the effects of the change are very difficult to forecast in the face of an industry which operates in periodic troughs in the boom/depression waves, where each company has its own method of operation. But the changes are real and often permanent and, hence, are openly operating alongside the traditional systems.

14.2 FACTORS INFLUENCING CHANGE

The factors influencing change are many, covering the whole spectrum of construction initiation, design, production and management, and include the aspects given below.

Size and technical complexity of construction projects

Construction projects are of constantly increasing technological complexity (with the mechanical services element accounting for approximately 40–50 per cent of the total cost). In addition, the requirements of construction clients are on the increase and, as a result, construction products (e.g. buildings) must now meet more exacting and varied performance standards (climate, rate of deterioration, maintenance and so on). Therefore, to ensure the adequacy of a client's brief which addresses the numerous complex client/user needs, a professional is now needed to evaluate the requirements in terms of activities and their interrelationship.

Expedited project delivery

As construction projects become more complex technically with clients' exacting requirements, clients now also demand that their construction projects be delivered earlier than before.

Design and production overlap

There is a tendency towards social levelling, an awareness of the need for cooperation on the part of the traditional professions, and a rejection of the 'them and us' attitudes and conflicts brought about by the longstanding separation of the design and construction functions. This has resulted in the overlap, at times, of the design and production functions and the use of specialist design consultants and contractors in the design and production of construction projects.

Influence of life-cycle costs

There is a greater awareness of life-cycle costs and their influence on the selection of suitable materials/components for construction projects. Most modern clients, at times, participate in the selection of building materials/components for their construction projects. Hence, information about the long-term effect of proposed materials/components on the total cost of a construction project is beneficial to clients when taking material selection decisions.

Clients' active participation

Corporate clients demand much greater involvement in matters (financial and otherwise) affecting their projects and sometimes have staff capable of performing this function. This active participation of knowledgeable clients is increasing the pressure for new, efficient and faster methods of construction procurement.

Funding arrangements and costs

There is an increasing sophistication in funding arrangements for construction projects. Furthermore, there is an overall reduction in the capital available from central and local government sources, and hence a resurgence of the private sector financing. As financing costs are escalating in a volatile financial market, and as modern clients constantly ask the question *'How much?'* in addition to *'How soon?'*, this requires a more efficient utilisation of funds and higher accountability.

Increased project participants

Design and construction have been bedevilled by user participation, several disciplines and multi-disciplinary teams (e.g. a multiple client/user who employ multiple design teams/specialist to design buildings to be built by multiple contactors and sub-contractors).

The input of an increased number of project participants requires an identification of those who possess the relevant construction information and the establishment of ways in which they might be encouraged to share the information with other participants. The CDM regulations now influence this. Furthermore, the increased inflow of project information has become complex and requires a new technique for its collection, processing, management and dissemination (information technology is now helping in this area).

Industrialisation of the construction industry

The industrialisation of the construction industry and consequential use of prefabricated components has effected a separation of the manufacturing and assembly processes and, hence, increased the demand for more dimensional accuracy than before. Moreover, the industrialisation has led to an increased wave of new building materials/components, and methods of construction need to be studied with specialist guidance in their selection and use. As a result of this, design has tended towards the selection and assembly of pre-designed and prefabricated components, materials and units from a multitude of manufacturers/suppliers, and their technical literature requires detailed study and investigation for component suitability.

Social restrictions

An increasing number of restrictions have been imposed by society on the methods of working and the environmental impact on construction projects. For this reason, designers are under pressure not to produce monuments unto themselves but, rather, should design projects under the watchful eyes of the client, producer, user and, at times, the observer or the general public.

14.3 RESPONSE TO CHANGE

As construction projects increase in size and complexity with corresponding increase in design time to allow all alternatives to be fully studied and specified, uncertainties increase. Hence, the ability to take rational design decisions in the circumstances is limited. In order to overcome the above external pressures of the traditional procurement method (Figure 14.1), the UK construction industry has devised what is known as an *integrated* construction procurement system (Figure 14.2). The integrated procurement system is a generic term for the many construction procurement systems which seek to integrate the design and production processes, while retaining the separation of responsibilities. It is a system which recognises that the design, manufacture and assembly processes of the construction product can no longer be left to chance, but must become one integrated process in order to deliver quality, value for money, speed and high productivity. Hence, it allows the contractor to play an active role in design as well as in production. This system may be considered under the following headings:

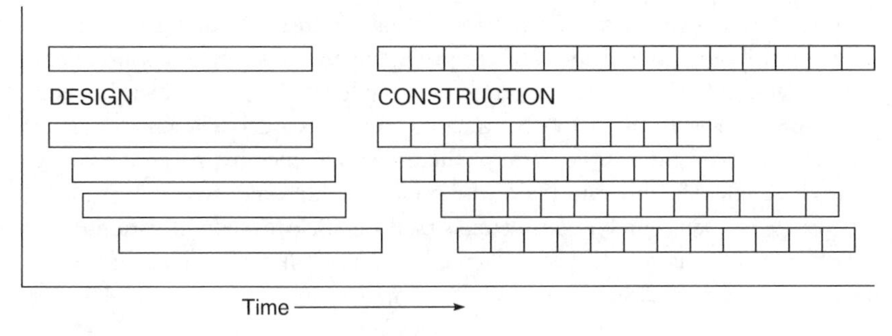

Fig. 14.1 Traditional procurement arrangement

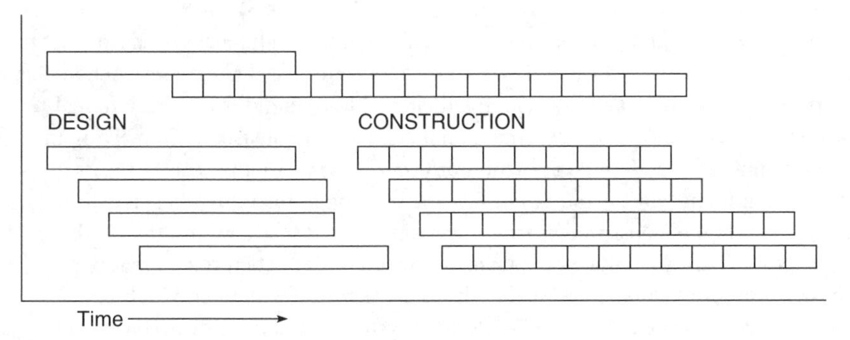

Fig. 14.2 Integrated procurement system (phased construction)

- Management contract
- Construction management
- Construction project management.

14.3.1 Management contract

Management contract is a contractual arrangement in which a client appoints a construction company termed the *management contractor* to manage and coordinate the design and production phases of a construction project. On appointment, the management contractor works alongside the client's professional advisers and this gives the client expertise not normally available at pre-contract stages of construction projects executed under the traditional procurement method discussed above.

The early selection of the management contractor enables that person to contribute to the design of the project. Although the management contractor does not directly design or construct any part of the permanent works, he or she gives expert advice on construction techniques and uses *on-site knowledge* to avoid the design of sections/elements that will be problematic to produce. Also, he or she seeks to meet the design requirements by the provision of specified common user and service facilities (e.g. tower crane, scaffolding, site offices, storage facilities, security), and letting each element of the project to a number of sub-contractors in work packages. The management contractor provides the necessary coordination and back-up services to the works contractors and is responsible for both the terms of their contract and the management of their works. Generally, he or she is responsible for the smooth running of the project within time, cost and quality parameters. The management contractor is paid for the provision of common user and service facilities in addition to an agreed fee based on a percentage of the estimated construction cost, for management input.

Growth of management contract

The management contract procurement method (Figure 14.3) has gained the modern client's acceptance and usage owing to the failure of the traditional system to meet their development needs. Because of their exposure to commercial risks, construction clients (particularly investment and development clients) have found that the traditional system of construction procurement is inefficient and unsatisfactory in meeting the speed required in modern construction project delivery. Another reason for its growth is the recognition of the evolution in the construction industry that has transferred the direct execution of construction from the main contractor to a multitude of sub-contractors. This has reduced the main contractor's production role to that of a construction production coordinator who takes charge of a number of sub-contractors on a construction site. Therefore, under this system, the contractor performs only facilitating and coordinating roles for payment.

Criteria for choosing a management contract

There are several procurement options open to the construction client who decides to build at present. However, the proposed construction project needs a careful assessment for the selection of an appropriate procurement path in the light of the

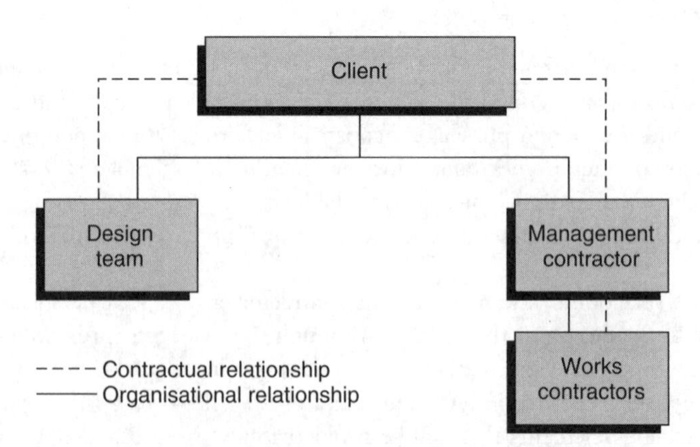

Fig. 14.3 Organisation structure for management contract

particular circumstances and constraints. The factors which may influence a decision to adopt a management contract procurement method may be summarised as follows.

1. *Early start on site:* Where time is of the essence and production cannot wait until design is fully defined and, hence, the production must proceed in parallel.
2. *Complexity of project requirements:* Where a construction project is complex technologically and therefore requires the input of a multiplicity of specialist designers and several sponsors/ultimate users with diverse requirements.
3. *Size:* Where the project is fairly large and, hence, needs complex organisation and coordination of a number of works contractors during production on site.
4. *Maximisation of competition:* Where the client requires maximum possible competition in respect of the price for production management expertise and resources.
5. *Early completion:* Where the client requires early completion and, therefore, there is the need for the adoption of a compressed and expedited development programme.
6. *Efficient use of resources:* Where the client wishes to capitalise on the efficient use of resources and his or her awareness that the contractor's advice on buildability at design phase achieves this objective.
7. *Lack of in-house management resources:* Where the client and his or her professional advisers have insufficient in-house management resources for the proposed construction project.
8. *Flexible budget:* Where the client has a flexible budget and can therefore take major risky financial decisions for the construction project.
9. *Contractor's expertise:* Where the contractor's production expertise is required at the design phase.

Appointment and range of services

Generally, management contractors are appointed after competition and interview. However, at times, the client may choose to dispense with the competition and approach a single construction company with which to enter into negotiations. When in competition, a management contractor's success depends on a number of factors, including:

- Level of management fee.
- Lump sum price for the provision of common user site facilities.
- Level of management expertise and experience.
- Degree of services offered within the management fee.
- Quality of support staff and equipment.

Depending on his or her terms of reference, the successful management contractor performs a range of functions at both design and production phases of the project. At the design phase he or she advises and assists the design team on various construction issues, which include:

- Buildability of design.
- Suitability of materials, specification, methods, forms of construction and programme.
- Availability of suitable materials, resources and works contractors.
- Programming and organisation of production.
- Information on work packaging, tendering arrangements and list of possible works contractors.
- Preparation of tender documentation and letting of direct contracts or work packages.

During the production phase, the management contractor continues the roles established at the design phase, and will also provide the following additional services during the projects production phase.

- Participation in the planning and programming of the various work packages.
- Coordinating, monitoring and controlling of the various work packages.
- Obtaining payment information from various works contractors and preparing cost/cash flow forecasts.
- Entering into contract/agreement with various works contractors.
- Providing, equipping, maintaining and removing common user site facilities.
- Coordinating, checking and making various interim and final payments to works contractors as and when certified by the client's professional advisers.
- Coordinating and controlling production; maintaining records of resource information and all contractual matters.
- Carrying out site inspection and implementation of safety, quality checking system, security and link communication system to facilitate progress.

The advantages and disadvantages of management contract procurement system are summarised as follows:

Advantages

- Early contractor involvement at the design phase may lead to better design and detailing which facilitates productivity and savings on production cost.
- The traditional design/construct split is eradicated, enabling the contractor to advise on quality, buildability, suitability and availability of labour, plant and materials and construction methods during the design phase.
- Design and construction are overlapped as well as overlapping the various work packages. This saves the overall project time.
- Risk of potential contractual claims is minimised as, from the project's onset, the managing contractor identifies contentious project information and recommends its modification prior to contract.
- The *them and us* attitude is eradicated as, from the project's onset, the management contractor becomes part of the project team working together to achieve a client's project objectives of time, cost and quality.
- Price is not the only criterion for selection of management contractor. Rather, in addition, the ability to make some technical and managerial contribution to the design and production of a project is also considered.

- Advance purchasing of essential materials/components and plant can be effected to ensure their availability for use when required.
- Early management contractor's appointment enables him or her to give information on the organisation of construction works, site layout, possible works contractors and tendering arrangements.
- Client obtains keener prices owing to increased competition for his or her construction project.

Disadvantages

- Problems of coordination between increased number of works contractors can lead to production delays and be grounds for contractual claims.
- Biased contract documentation may be drawn up which unfairly allocates responsibilities to works contractors who may not be well equipped to perform.
- Client pays twice for the duplication of site services, attendances and general preliminaries.
- Owing to financial pressure, works contractors may be forced to forgo some of their proper entitlements under the contract.
- Client suffers financial loss for making good defects if a works contractor fails to remedy faulty workmanship.
- Client is not aware of the financial commitment before the commencement of a project on site.
- In the event of the termination of a works contractor's contract (due to non-performance), the client bears the cost of arranging the completion contract.
- Client may have no redress against a management contractor in respect of performance and quality, works contractor's work, late completion and recovery of damages for late completion.
- Management contractors are (sometimes) pressurised to accept some of the financial risks inherent in the production such as maintaining satisfactory performance and quality of work, time overruns, latent defects and design failures.
- Client has two sets of fees to pay for the professional services of design and construction management.

Generally

The management contract procurement method makes heavy demands on works contractors. They are involved with some aspects of building design and, hence, are expected to coordinate closely with the project team. Their respective inputs are also considered vital to the successful completion of a construction project which adopts the management contract procurement method. Furthermore, under the management contract procurement method, the contractor is responsible for the selection of the works contractors but allows the client and his or her professional advisers to be involved in the selection process. In principle, the selection process is very similar to that of the traditional system, but under the integrated method the management contractor arranges and makes all the selection of works contractors for the various works packages. In addition, the works contractors normally work closely with the management contractor and therefore, the management contractor's selection is confined to those works contractors of known ability and experience.

While the roles of the client's professional advisers will differ in some respects from those in the traditional system, they are still responsible for the design and cost management of the project. During the production phase, the architect supervises the works and provide all additional production information required. The quantity surveyor will be responsible for the preparation of contract documentation for the various work packages, cost reports, final accounts and the evaluation and settlement of all contractual claims.

14.3.2 Construction management

Construction management is one of the procurement options for the delivery of a high-quality construction product on time at the least cost. It is similar to the management contract approach but has the following exceptions:

- The construction manager is a consultant who is employed by the client in a purely managerial capacity; therefore he or she does not accept any liability for non-completion unless resulting from professional negligence.
- There is a direct contractual relationship between the various works contractors and the client.
- The design team and construction manager possess equal professional status.
- In this method, the client has direct responsibility for his or her project and flexibility in procuring the various specialist services.
- The client assumes a greater share of the financial risks owing to his or her active and continuous involvement and also as a result of entering into direct contracts with the works contractors.

Emergence of construction management

The construction management approach to construction procurement emerged as a result of the construction client's demand for better quality and faster production at lower costs. The factors that contributed to its demand include the high cost of finance (especially during the production phase) and the need for an early return on investment. Another recent factor contributing to the demand for adoption of a construction management procurement method is the greater awareness of the total life-cycle costs of construction projects. As the traditional procurement system has failed to respond adequately to the current needs of construction, the construction management route has emerged as an alternative procurement which has a lot to offer clients in their development aims. The organisational relationships for construction management are shown in Figure 14.4.

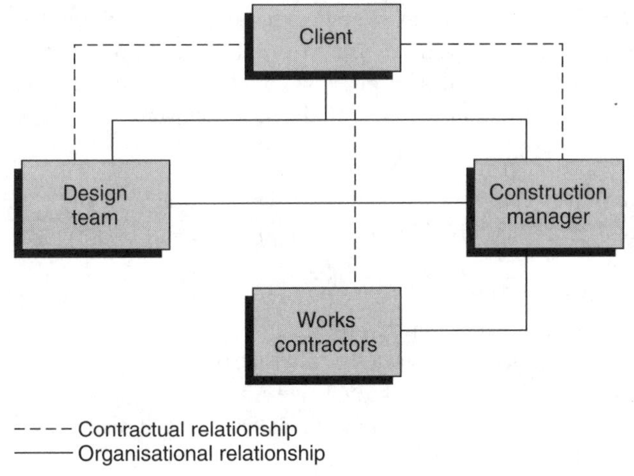

Fig. 14.4 Organisational structure for construction management

Criteria for choosing construction management

The criteria for choosing the construction management procurement path do not differ from those for the management contract construction procurement method. However, the main factors that influence its adoption include the following:

- Early completion of the construction project is required by a client.
- The project is of fairly large size and is technologically complex.
- The construction project has complicated requirements.
- Design and production phases must overlap to expedite production.
- Input comes from several sponsors/ultimate users with diverse requirements.
- The requirement of a maximum price.
- The demand of an effective utilisation of production resources.
- A need for management input in the design phase.

Appointment and range of services

Unless the construction manager has been obtained in-house (i.e from the client's own establishment), he or she is usually appointed after competition and interview. However, like the management contract procurement approach, the client may choose to dispense with the competition and, instead, enter into negotiations with a single professional leading to the appointment. In the competition, the construction manager's success normally depends on a number of factors, including:

- Level of management fee.
- Services offered within the management fee.
- Level of construction manager's expertise and experience.
- Quality of support staff and equipment.

Depending on his or her terms of reference the successful construction manager performs the following functions at both pre-contract and production phases of the construction project. At the design phase, he or she advises and assists the design team on various construction issues, including the following:

- Buildability of design.
- Coordination between drawings and also, between drawings and specifications.
- Suitability of materials, specification, methods, forms of construction and programme.
- Availability of materials, labour and other resources.
- Programming and organisation of production.
- Production cost information on design alternatives.
- Identification and definition of the separate work packages.
- Establishment of temporary on site facilities for client, designers, quality inspectors and works contractors.

During the production phase, the construction manager continues the roles established at the design phase, in addition to the following:

- Planning, programming, controlling and coordinating the project and the interface of all participants on site activities.
- Keeping to the project cost plan and programme through anticipation rather than reaction, and feeding real cost information to the design team.
- Taking responsibility for the time/cost relationship and ensuring that the project progresses in line with the client's development strategy.
- Establishing effective lines of communication to enable free flow of project information between all participants.
- Controlling the procurement of production resources to meet the project's programme.
- Interpretation of drawings and specification and the identification of necessary amendments.
- Carrying out periodic site inspection and implementing safety, quality checking and security systems.

Generally

At the design phase of the project, the construction manager draws up a list of prospective tenderers. Tenderers for each work package are interviewed and, thereafter, a short list of tenderers is prepared from whom competitive bids are sought. On return of tenders, they are evaluated by the client's quantity surveyor and recommendation is made to the client who places an order with the successful tenderer through the architect or project manager.

During the progress of the works, the architect performs the supervisory function and provides all production information. The client's quantity surveyor, on the other hand, will carry out the normal contractual duties of preparing contract documentation for the various work packages, cost reports, interim valuations and final accounts. In addition, he or she will have delegated responsibility for the ascertainment and settlement of all contractual claims.

The advantages and disadvantages of construction management procurement methods may be summarised as follows:

Advantages

- The construction manager provides the design team with positive but disinterested construction production advice at the design phase.
- The construction manager's skill in communication, organisation, leadership and team building enables a completed construction product of high quality to be obtained on time.
- The good contractual relationship between the client, the design team, the construction manager and the works contractor which this method offers enables the client to exercise some control and influence over production.
- The construction manager's early involvement in design enables him or her to advise on production of buildable designs and also to identify conflicts between parties, and to review drawings and the specification.
- The overlapping of design and production phases can lead to accelerated production and early completion of the project.
- Advanced orders of items on long delivery (which will be needed early in production) can be arranged.
- The construction manager's ability to encourage production of buildable designs can improve productivity and so reduce costs.

Disadvantages

- While a budget can be set up in the multi-packaging of the work, it is not possible to obtain a total price on the project until the last bid package is awarded.
- The client is exposed to a high degree of risk as the construction manager does not take responsibility for late completion, faulty workmanship and so on.
- The client risks the cost of abortive work as the project may not be fully planned or developed prior to commencement of production.
- Failure to manage and coordinate work package interfaces adequately can be the cause of construction disputes and contractual claims.
- The client pays more in professional fees.
- Works contractors are subjected to the provision of some site facilities which are normally supplied by the main contractor under a conventional contract.

14.3.3 Construction project management

Construction project management is a management function in construction design and production which comprises the planning, costing, evaluating and controlling construction projects on behalf of the client, so that it is completed on time, to specification and within budget. Under this system, the client appoints a construction professional who is given the responsibility for various functions of site identification, land assembly, approval, funding, design, construction and marketing. This construction professional so appointed is known as the 'project manager' and unless he or she is the client's employee, is paid an agreed fee based on a percentage of the total cost of the construction project. Generally, the project manager is the bridge between the client and the project team and is responsible for the task of helping to establish the overall project objectives and directing, controlling, planning and coordinating the efforts of the project team to the achievement of those objectives.

It must be mentioned, however, that construction project management is not a construction procurement system. The role of the project manager is to select the most appropriate procurement system for the project. Construction project management is therefore concerned with the management of the whole construction project (i.e. the translation of the client's needs into a finished functioning physical product). Thus, project management covers (a) the coordination of the diverse interests of ultimate users (of the completed facility) and the client in order to formulate a unified client requirement and (b) the organisation of the process of design and construction of the project to comply with the client's requirements economically and timely.

Emergence and growth of construction project management

Construction project management has emerged as a result of complex forms of construction required by construction clients over the last few decades. Modern buildings, for example, have complicated air-conditioning systems, electronic services, surveillance systems and security systems. Also, the pressure of commercial sophistication is forever requiring expedited construction project completion through greater integration between design and production phases. As construction project management ensures that the design, manufacture and assembly processes of building are no longer segregated but become an integrated process to deliver quality, value for money, speed and high productivity, most clients find it advantageous to employ a project manager.

Criteria for adopting construction project management

Generally, all construction projects vary; nevertheless, the combination of time, value, people, place and so forth is certain to be unique in all projects. Hence, the criteria for adopting construction project management depends on the variation of unique factors, including the following:

- High value
- High complexity/risk
- A short timescale for delivery
- Resource scarcity
- Larger number of participants
- Geographical dispersion of site.

Appointment and range of services

As a construction professional, the project manager may be appointed through either competition, negotiation or recommendation. Like the construction manager, his or her success in competition normally depends on the following:

- Level of management fee.
- Degree of services offered within the management fee.
- Level of management expertise and experience.
- Quality of support staff and equipment.

Depending on the exact terms of reference and full extent of authority, the successful project manager will perform the following functions at both the pre-contract and production stages of a construction project. At the pre-contract phase, the project manager's role normally includes:

- Appraising the client's requirements and assessing the general form of development, location, availability of funds, time constraints, etc.
- Establishing of the feasibility of the project, taking into account objectives, market trends, financial and technical factors, planning approval, timing, views of interested parties, etc.
- Recommending the selection and appointment of designers and other consultants.
- Formulating the design brief and selecting project particulars.
- Drawing up an action plan showing time and sequence of activities, management structure, times and sequence of communication, arrangement with key suppliers and other project participants, all compatible with the client's requirements.
- Arranging for the preparation and submission of preliminary drawings for planning approvals and the establishment of preliminary budget.
- Preparing a feasibility report outlining all aspects of the construction project for the client's approval.
- Ensuring that the design consultant's design proposals, development and specifications are properly coordinated and meet the client's and the planning authorities' approval.

During the production phase, the project manager's role will still be managing the project in accordance with the client's brief, and in the process his or her function will include:

- Establishing the preferred procedure for preparation of contract documentation, tendering and selection of contractor/sub-contractors.
- Evaluating tenders received and recommending contractor/sub-contractor for client's approval and appointment.
- Monitoring the production progress and the effects on the budget and programme as a result of variations or delays.
- Motivating project participants, encouraging free flow of construction project information and seeking and nullifying any potential items of dispute.

- Ensuring that the contractor/sub-contractors receive periodic payments for work satisfactorily completed and that their final accounts are dealt with in accordance with the contract conditions.
- Ensuring that the completed development complies with the client's original requirements and checking that all plant and equipment are functioning properly before they are commissioned.
- Assisting the client in carrying out commissioning and making sure that he or she obtains as-built drawings and operation and maintenance manuals. In addition, ensuring that all defects have been remedied before release of final payment to the contractor and/or sub-contractors.

Attributes of a project manager

Generally, construction project management concentrates on project leadership normally exercised by one person – the project manager – who is charged with the responsibility of making sure that the project is conducted and implemented according to established goals, directives and plans. Moreover, managing a construction project successfully today demands a number of skills and knowledge necessary to control time, cost, quality and performance. The project manager, therefore, possesses the ability, skill and expertise in welding together an effective team for a timely project delivery. An ideal project manager who can successfully implement a change, coordinate all resources and manage all disciplines to complete a project should possess several skills, which include the following:

1. *Organisational ability:* The successful project manager must have an organisational ability and also be able to manage the construction project, its numerous participants and the sundry problems that may arise.
2. *Broad experience:* While a project manager may come from one of the construction disciplines (e.g. architecture, quantity surveying, engineering or general building) he or she must be familiar with all aspects of construction design and production to be able to identify problems and take the necessary measures quickly to address them.
3. *Communication skills:* A project manager must be able to communicate effectively both verbally and in writing. Moreover, he or she must also have the skill of listening to others in order to deal with day-to-day problems that may arise.
4. *Leadership respectability:* A project manager must be able to command respect and win the confidence of all persons associated with the construction project.
5. *Effective decision making:* The project manager, being a leader of a project team, must be able to analyse situations and take decisions which will drive the project smoothly to a successful conclusion.
6. *Motivation:* The project manager's ability to motivate and create a good atmosphere for team work within a project organisation has a crucial bearing on the successful completion of a construction project.

7. *Effective delegation:* The ability to delegate skilfully is an important attribute of a project manager. The knowledge and expertise required for the design, production and management of a modern construction project is too diverse for any one person to attempt to acquire them all. The project manager, therefore, delegates responsibilities to others and is therefore able to obtain the necessary information required for driving the project to successful completion.

In addition to the above management skills, the project manager must be able to display independence and impartiality, cost awareness and understanding of human relationships. It is also essential to have the ability to lead a team of experts, command respect and win their confidence.

The advantages and disadvantages of construction project management are summarised as follows:

Advantages

- The client's brief is prepared by a skilled construction professional who has the experience in the production of an adequate briefing document.
- The 'them and us' confrontation is averted as the project manager welds all participants into an effective construction project team and ensures that all are pulling in one direction.
- The client benefits from having just one person to deal with in all matters concerning the construction project.
- The architect is released from the tasks and problems associated with managing the construction project and is therefore able to concentrate on design matters.
- The project manager is able to keep the balance between aesthetics, functional use and cost, and also ensures that the client is aware of the financial effects of any variations to the original scheme.
- The project manager performs an independent and disinterested role and uses this position well in directing, coordinating and solving construction problems and disputes as and when they arise.
- Design can be tailored to overlap production, thereby reducing the overall project time.
- The client can be sure that his or her interests are being safeguarded by the project manager who is able to provide a skilled, unbiased interpretation of the requirement proposals made by the various consultants.

Disadvantages

- The client's project cost increases as an individual is required within his or her organisation to act as a focal point to ensure that the client's requirements are being met.
- The project manager requires executive authority to achieve high performance, but this authority may be lacking.
- The project manager carries no financial risk and can only be sued for damages if professional negligence or breach of contract is established.
- The client incurs an additional cost as a result of the project manager's fees.

Generally

As changes in technology and market forces are demanding greater speed and economy in construction project delivery, all the above are being adopted by construction clients, construction professionals and major construction companies in the UK as a means of addressing the needs of the ever-changing scene. When the integrated procurement system, is compared to that of the traditional system, the following differences can be noted:

- Design and production phases are overlapped and at times the design and production phases of work packages are also overlapped.
- The architect loses the traditional role as a project team leader. This role is taken over by a construction professional known as a 'project manager'.
- The main contractor does not directly take part in any of the construction work. Instead, he or she provides the common user facilities and manages and coordinates the activities of works contractors.
- Specialist construction managers are, at times, engaged to advise on buildable designs and manage the production on site (instead of the contractor).
- The management contractor's expertise in buildability, availability of specific materials, plant and site operatives is utilised and, therefore, has an input in both design and production phases of the project.
- To assess contractor's management expertise and production skills, a two-stage tendering system is adopted for the selection of a management contractor.
- In the management contract system, the management contractor is responsible for selecting all works contractors.
- The problem arising from a division of management responsibilities is overcome and, as a result, the client enjoys the benefits of having a single point of contact.

14.4 A TYPICAL INTEGRATED PROCUREMENT SYSTEM IN PRACTICE

14.4.1 Project initiation

Like the traditional construction procurement system and all construction projects, the client is the initiator of the construction process. The client normally begins with the appointment of project manager (instead of an architect) and briefs the project manager of the project's requirements.

14.4.2 Project leader

Under this approach the project leader is the project manager whose role consists of advising the client and organising, coordinating, controlling and supervising the project. To perform this role effectively, the project manager must have the ability to organise and create a good working environment, weld the project team together, motivate them and ensure that all are working in unison.

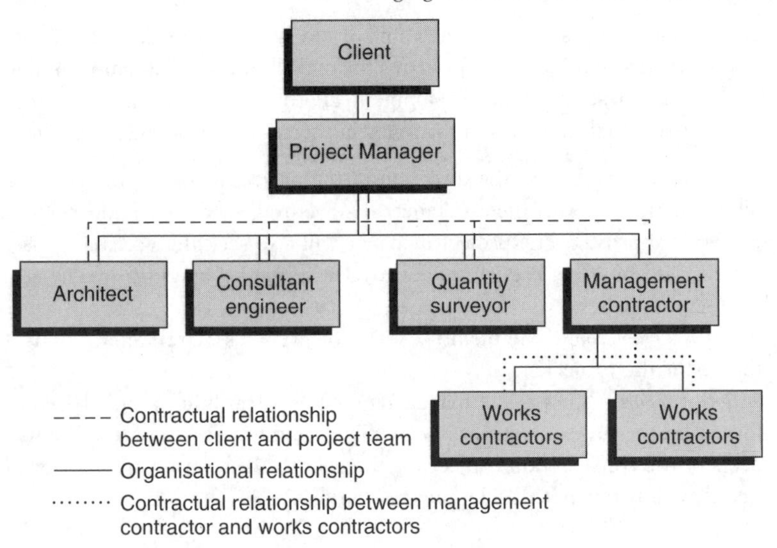

Fig. 14.5 Simplified organisation structure for a typical integrated management system

14.4.3 Organisation structure

The traditional method of managing construction projects separates design and production functions and also aims to place the major production risk on the contractor. Conversely – designed to meet the challenges, the complexities and the ever-accelerating pace of the development process – the integrated management system transfers most of the production risks to the client. It is the client's awareness of these risks that makes the selection of a capable project team crucial to the achievement of the development goals. Therefore, the fielding of an effective project team reduces the risks by simplification of individual work package requirements, reduction of procurement time, and bringing forward the financial returns. A typical organisation structure for the project team can be seen in Figure 14.5.

14.4.4 Pre-contract project organisation: feasibility stage

On appointment, and after briefing, the project manager starts working on the client's requirements. In order to complete the project expeditiously, after the initial investigations and conviction of the feasibility and viability for the client's project, the project manager will assist the client in the formulation of a clear project brief. The project manager also advises the client of a need to appoint a design team, composed of an architect, a quantity surveyor and consultant engineers. On their appointment, the project manager briefs them on the specification of the client's construction project. He or she will review the project specification to ensure a full understanding of the project requirements by all participants. Once the project manager is satisfied that the project objectives are fully understood, the project participants develop the strategy for achieving success in the project's delivery. This move clears the way for the project manager to establish lines of communication, extent and frequency of meeting of all the participants. In addition, he or she prepares the overall project programme in line with the client's total timescale for the development, and the participants are expected to perform their required functions in the following stages of development activities:

1. *Preliminary design:* Under the direction of the project manager, the architect designs the project with the assistance of structural/services engineers to meet the client's approval. The project manager will also seek the necessary basic planning, and building regulation approval will also be sought from the area planning authorities.
2. *Preliminary budgeting:* The quantity surveyor prepares a preliminary budget estimate for the project.
3. *Feasibility report:* The project manager prepares a feasibility report based on information the project team has been able to gather. This report will outline all aspects of the construction project for the client's decision on whether or not to proceed with his or her development ambitions.

14.4.5 The management contractor

The management contractor is appointed when the client has given the all clear signal after the feasibility stage, but before detail design and subdividing of project into work packages. Normally, the method of management contractor selection is by 'two-stage tendering'. However, all works contractors are selected under the provisions of single-stage selective tendering (see Chapter 5 for discussion of both approaches to contractor selection).

14.4.6 Planning and programming

After the project manager has drawn up and received approval of the client's brief, the project team sets parameters and goals for overall project planning. Items covered in the programming at this stage include:

- Scheme design.
- Preparation of contract documentation work packages.
- Interview, selection and appointment of works contractors.
- Preparation of design working drawing.
- Production of each work package.
- Key dates for specific activities highlighted for:
  - Sending work package documents out to tender.
  - Award of work package to works contractors.
  - Work package completions.

Within individual work packages, 'milestone' dates for completion of critical interface works may be necessary, and to achieve the project as planned the project manager will, on implementation, monitor the programme and institute measures to correct any deviations as necessary.

14.4.7 Scheme design and contract documentation

The architect carries out the normal design function at the scheme design phase of the project. However, under the integrated management arrangement, the managing contractor is always available to provide the architect with advice on buildable design, suitability and availability of material/components. The project manager will also ensure that the design is well managed and completed on time, within budget and contains all the client's defined requirements. In addition, at this stage, the project manager will identify for appointment at the appropriate point in the design process, specialist contractors whose design skills will be relied upon.

When the scheme design is completed and the project has been subdivided into various work packages, the quantity surveyor sets a budget in the cost plan against each of the work packages; following that, he or she will measure the available work package drawings and produce contract documentation on which to invite tenders. The project manager double checks the contract documentation to ensure that it contains all the project's defined requirements. In addition, the contract documentation will be examined with the aim of uncovering and eradicating any open ended issues, loose ends or vagueness in the contract documentation that could later surface and cause problems.

14.4.8 Works contractors

Most works contractors are selected by either competitive tender or (in some cases) by negotiation based on either bills of quantities or a work schedule with preliminaries and specifications. However, sometimes a performance specification is also adopted. Each tender is evaluated and recommended by the quantity surveyor before the works contractor is engaged on the project.

14.4.9 Post-contract project organisation

Post-contract project organisation comprises the following functions:

Supervision and coordination on site

Besides the pre-contract coordination aimed at achieving a tender action at a date consistent with the agreed construction programme, the management contractor produces and regularly updates a status schedule for all work packages. The management contractor is responsible for ensuring that any revision to the status schedule is communicated to the project team and that all the essential activities leading up to tendering for the various work packages are under control. Also, the management contractor is responsible for on- and off-site supervision of the works contractors, programming, resources management, expediting and provision and maintenance of common user services. Moreover, it is his or her responsibility to resolve all production problems to do with interface between activities whenever they arise.

The architect, on the other hand, performs the usual site inspections to ensure that the project is progressing as planned. He or she may issue instructions to clarify or amplify information already given to the works contractors but must seek the authority of the project manager to issue any instruction which can affect both cost and time. In addition, the architect will carry out any design modifications necessary and authorised by the project manager. He or she will issue certificates for interim payments and all other certificates such as certificate for non-completion or practical completion and carry out practical completion inspections.

The project manager monitors and controls progress on site; arranges meetings regularly to ensure that everyone is pulling in the same direction in order to avoid delays; and will also need to monitor the cost effect of lost time on the programme and the budget as a result of variations. In addition, he or she must establish whether a saving is being made or extra expenditure is being incurred in each work package and introduce immediate corrective measures to remedy the situation if necessary.

Cost management

As tenders for the work packages are received and evaluated, the accepted tenders must be reconciled with the budget. Periodically (at least monthly) the client should be advised on the budget status, which takes into account the following:

- Additional cost or savings identified during design but yet to be reflected in the work package tender sum.
- Total value of all authorised variations with a summary of the principal items.
- Estimated current and anticipated effect of cost inflation.
- Estimated cost effects of current and anticipated delays or disruptions.

As his or her normal post-contract duties, the quantity surveyor is responsible for the preparation of interim valuations and the final account. The quantity surveyor will also have delegated responsibility for the ascertainment and settlement of all contractual claims.

Commissioning

The architect and project manager assist the client in carrying out the commissioning and participate in carrying out all equipment performance tests. They will see to it that all operating manuals, maintenance manuals, as-built drawings and so on are supplied to the client. Moreover, prior to the release of retention, they ensure that all remedial works have been carried out and that everything is functioning to the client's satisfaction.

SUMMARY

In this chapter we have discussed the ever-changing construction procurement scene. Commercial pressures to succeed which influence clients and contractors to complete buildings more quickly is not becoming less intense. Moreover, the increased size of projects, coupled with the increased number of participants and the expansion of production information and use, do not signify that the integrated construction procurement techniques have been ideally devised to deal with the deficiencies in the traditional construction procurement process. While the integrated system has been used on some projects and some successes have been recorded, it has not been able to solve all the ills in construction procurement and management. It has not been able to resolve, for example, the inadequacy of briefing and construction information, lateness in the issue of construction information and clarity of communication between client, designer and contactor. For this reason, there is always a mis-match between the construction client's expectation and reality due to inadequate briefing and communication procedures.

Instead of tackling this fundamental problem, the integrated system seems to create further divisions of functions such as specialists taking and developing client's brief, performing design consultancy, managing design and planning/ programming design and construction. One can say that many of these divisions are often the result of natural progression; nevertheless, the above functions were once the birthright of the architect. Furthermore, under the same innovative moves, such terms as project manager, client's representatives and so on have appeared on the construction scene.

FURTHER READING

Barnes, M. 1988 *Construction project management*, Vol. 6. Butterworth, May, pp. 69–79.

Birnberg, H. 1993 Roles of a project manager. *Architecture*, July, pp. 115–17.

CIOB. 1988 *Project management in building*. Chartered Institute of Building.

Hayes, R. 1985 The risks of management contracting. *Chartered Quantity Surveyor*, December, pp. 197–8.

Hibberd, P.R. 1990 Contractors design liability under the standard forms of building contract. *CIOB Technical Information Service, No. 118.*

Hughes, W.P. 1991 An analysis of construction management contracts. *CIOB Technical Information Service, No. 135.*

Jones, E. 1993 Construction management and project management – the differences in structural and its impact on project participants. *International Construction Law Review*, pp. 348–65.

Lock, D. 1991 *Project management.* Gower.

Reiss, G. 1992 *Project management demystified.* E&FN Spon.

Snowdon, M. 1980 Project management. *The Quantity Surveyor*, January, pp. 2–4.

Walker, A. 1984 *Project management in construction.* Granada.

Thompson, P. 1991 The client's role in the project management. *Project Management*, **9** (2), May, pp. 90–2.

CHAPTER 15

GROWING NON-TRADITIONAL SYSTEMS

15.1 INTRODUCTION

Like the integrated construction procurement system, which sprung up out of sheer failure of the traditional system to respond effectively to the needs of the modern construction client, other non-traditional construction procurement systems have been introduced along the lines of the integrated system. These systems are many, but design and build, British Property Federation (BPF) and fast track systems have been identified for consideration in this chapter. Like the integrated construction procurement system, all the above-named systems have contrived to bring the activities of design and production closer together and have managed, in some respect, to reduce project delivery times. They offer a single point of contact for the client's development aims and dispense with much of the administrative time and cost inherent in the traditional modes of construction procurement. In addition, they do also recognise the important role a modern building contractor can play in the process of a building design, production and management. For this reason, the building contractor plays an active and recognisable role in all the above named construction procurement systems.

15.2 DESIGN AND BUILD

The design and build method of procurement enables one building contractor or a construction company to take full responsibility and carry sole liability for the design and construction of a building (Figure 15.1). Hence, in effect, apart from the construction client's role, design and build essentially combines all the fundamental tasks in construction project design, production and management in a single package. Therefore, after identifying the need for a building, the client states his or her requirement adequately in terms of physical design needs as well as the intended physical use. A selected number of building contractors are invited to submit their proposals together with estimated costs. The system invokes design competition, which is absent from other construction procurement systems and

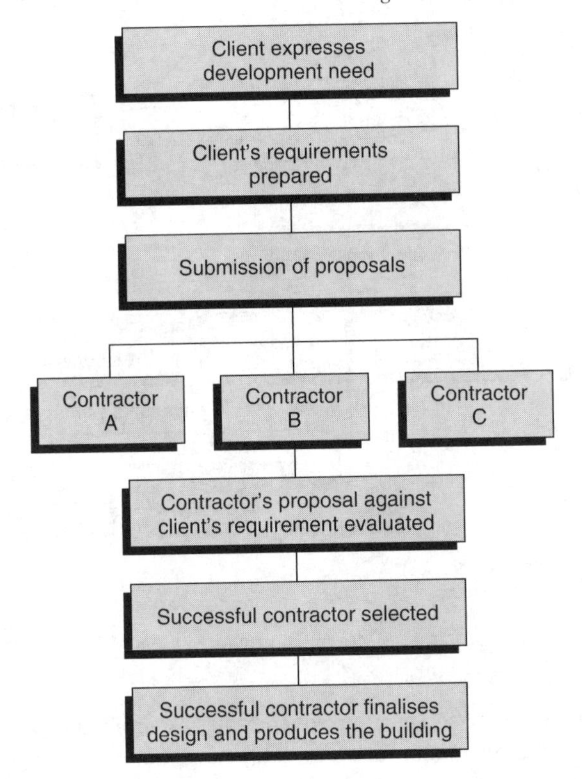

Fig. 15.1 The design and build process

permits the optimisation of design and production costs. This construction procurement method is suitable for standard buildings, industrialised systems such as factories and warehouses, office buildings, residential buildings, educational buildings, hotels and so forth. On large, complex or specialist projects, design and build companies may decide to appoint a designer from consultancy firms. In such a case, the appointed designer's responsibility is to the design and build company and not directly to the employer (see Figure 15.2). However, the client normally appoints agents (architect, quantity surveyor, engineers) to look after his or her interest and also to ensure that the contractor's proposal receives planning approval.

15.2.1 Growth of design and build

Although the design and build construction procurement system has been in existence for many decades, its use came to prominence within the UK construction scene in the late 1970s. In the UK today, the building contractor led design and production option is the fastest growing sector in the construction industry and it is claimed to be enjoying some 15–20 per cent of the construction market. Its rapid growth has been the result of the client's belief that responsibility

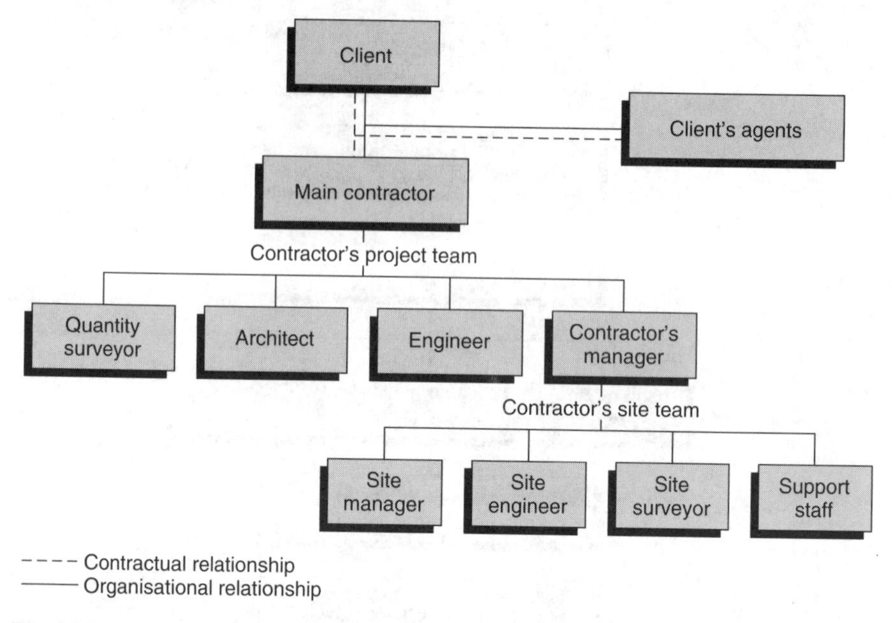

Fig. 15.2 Organisation structure of design and build contract

for design should not be so far removed from the responsibility for production. The need for better integration between design and production has led to the client's recognition that design and build is an appropriate answer for some large and complex building and civil engineering projects. Moreover, there is general agreement that a closer association between design and production reduces the incidence of misunderstanding, improves communication, promotes production of buildable designs, and makes best use of a contractor's resources and knowledge of construction project production. However, the effectiveness of such a closer association requires a far greater involvement of, and commitment by, the contractor at the early stages of a project, and design and build is one form of construction procurement system that meets this requirement or objective.

15.2.2 Criteria for choosing design and build option

From the point of view of his or her property development ambition, the client will always choose a construction procurement option which best suits the priorities. However, while a desire to procure for less money and less utilisation of resources may influence a client's choice of an appropriate construction procurement route, this will not be the final deciding factor. The main characteristics under which the design and build procurement option may be appropriate for a client's construction project are when:

- Early delivery of project is required.
- Project is of technical complexity.

- There is the need for early start on site.
- Price certainty (usually fixed lump sum) is required prior to production.
- Construction project is prestigious and, hence, single point responsibility is required.
- Economy (time, cost, function, quality, value for money) is required.

15.2.3 Design and build in practice

Generally, the services provided by design and build contractors are diverse; however, a common feature is that the contractor is responsible for not only the construction project's design but also for its production. The design and build process is set in motion when a client approaches a building contractor with a set of requirements defining what is to be constructed. Alternatively, the client may introduce an element of competition in the contractor's selection procedure by inviting a list of contractors with the necessary expertise to submit competitive proposals based on his or her requirements. The extent and details of a client's requirement vary from construction project to construction project, but this may be expressed in terms of either performance criteria or plain specification. It may also be merely a brief statement as to the type of building, use, floor area, number of storeys and so on. In each case, however, the client's brief must be precise to allow for its development, costing and evaluation.

15.2.4 Contractor selection method

The method of contractor selection under the design and build construction procurement route is essentially a single-stage selective tendering method, as can be seen in Figure 15.1. Nevertheless, this does not preclude any negotiations with the successful building contractor to amend details and specification or to consider any alternatives he or she may offer. In the process of tendering, each tenderer is required to provide, with the tender, sufficient details and typical working drawings to facilitate a full technical evaluation by the client's agents.

15.2.5 The tender documentation

A typical tender documentation for a design and build project would include the following:

- Invitation to tender.
- Site survey, site layout drawing delineating the extent of external works, roads, paths, sewers. etc.
- Design brief, setting out in detail what is to be built, the tenderer's areas of design responsibility and any constraints upon the design.
- Preliminaries and conditions of contract.
- Tender specification for substructure, superstructure, mechanical and electrical installation, etc.

- Pricing schedules and main summary.
- Form of tender.

15.2.6 Contractors' proposals

Each contractor's proposals normally consist of working drawings, written information, structural calculations of members, construction programme and estimated cost. These proposals should be presented in a way that facilitates evaluation and/or checking by the client's agent.

The design proposals to the client's specification, the structural calculations and the estimated cost are carried out by the contractor's in-house design team of architects, engineers and quantity surveyors. Where a bidding building contractor has no such in-house facility, his or her quantity surveyor prepares a detailed brief and appoints private consultants to carry out the design, structural calculations as well as the provision of cost estimates. However, whether or not the proposal is carried out in-house, it should be able to receive planning approval, be competitive, and meet the client's requirements in terms of function, appearance, space and cost. On evaluation, the contractor whose proposals most closely match the client's requirements is awarded the contract. The successful contractor commences production on site when the contract is signed and the client has issued a formal order to commence production.

15.2.7 Management of production

During production on site, the contractor's site manager performs the usual site supervision functions. The contractor's architect and engineer conduct periodic site inspections to ensure that the specified quality is being met. During the progress of site production, the contractor's architect will hold periodic site meetings which the client's agents are entitled to attend. The contractor's quantity surveyor provides cost advice on alternative forms of construction; prepares bills of quantities both as a basis of tender and sub-contract enquiries; and accepts full responsibility for the accuracy of the tender documentation. In addition, during production, he or she controls production costs, provides cash flow forecasts, prepares internal cost reports and cost value reconciliation and advises as to the anticipated final cost. In addition, the contractor's quantity surveyor will agree all valuations for interim certificates, valuation of variations and final account with the client's agent (the quantity surveyor.)

15.2.8 The role of client's agents

Under the design and build construction procurement route, the client may still need advice from construction professionals on the formulation of the project brief, functional requirements, structural stability and cost evaluation of a contractor's proposals, financial management and quality control. Therefore, the client requires the services of an architect, a structural engineer, a services engineer and a quantity surveyor who have the appropriate skills to check and supervise the various stages of the development project. Hence, the role of the client's agents is to monitor and control the project on behalf of the client in order that the client may achieve his or her development aims, and this role is performed as follows:

Pre-contract phase

Architect The architect assists the client in the formulation of an appropriate construction project brief on which competing contractors will base their proposals. This requirement may be based on functional, space and cost criteria and, on receipt of proposals, the architect evaluates the functional, space and aesthetic criteria of various proposals and submits his or her findings and recommendations to the client for a decision.

Structural engineer The structural engineer checks the structural stability and/or calculations of the contractor's selected proposals and submits a report of findings, with recommendations to the client.

Services engineer The services engineer checks the quality of services equipment contained in the contractor's proposals and their various locations in the building and submits a report of findings, with recommendations to the client.

Quantity surveyor The quantity surveyor advises the client on current building costs and the relationship between capital expenditure and maintenance costs on the preparation of the original brief. He or she also ensures that the specification defining the scope and character of the project is comprehensive (to avoid unforeseen extra costs) and that the contract provides a proper basis for valuing variations during the course of the contract. On receipt of contractors' proposals, the client's quantity surveyor produces a comparable estimate of contractors' proposals, for the assessment of the merits of each particular offer. After the foregoing exercise, the quantity surveyor is able to recommend the most favourable proposal to the client.

Production phase

Architect The architect's role during the production phase consists of periodic site inspections to ensure that the quality specified is being met and that good construction methods are being practised. The architect and the building contractor will carry out a final inspection before commissioning. It is also the architect's responsibility to ensure that the client obtains all relevant documents such as as-built drawings, operation and maintenance manuals at the commissioning stage.

Structural engineer The structural engineer's role during the production phase consists of periodic site inspections and checking to ensure that the specified materials and right sections of members are being used on the project. In addition, he or she may give professional advice on any structural changes deemed necessary.

Services engineer The services engineer checks and tests the various services equipment and their installation, and also assesses their suitability for their intended purpose.

Quantity surveyor The client's quantity surveyor's role during the production phase is the preparation and agreement of periodic valuations for interim payments with the contractor's quantity surveyor. In addition, he or she negotiates proper value for variations from the original specification and/or drawings that may be required by the client and agrees the final payment to the contractor.

Advantages and disadvantages

The advantages and disadvantages of the design and build construction procurement method are summarised as follows:

Advantages

- The integrated design and construction allows for design and management input from the building contractor, and this leads to production efficiency in terms of cost and time.
- The method allows for a simplified contractual arrangement between client and contractor with a single-point responsibility and improved communication channels between parties to contract.
- Project duration is shortened due to contractor's familiarity with his or her system and parallel working on design and construction.
- Client's total financial commitment is known at an early stage and, provided the client does not introduce major alterations, this will not change.
- Client obtains competition in design as well as in price.
- The closer contractor/client relationship leads to a more efficient design.
- Client obtains a design cost element lower than that which an independent designer would charge under other methods.
- Construction projects using the system have the potential for early completion and lower overall costs.
- Innovation in construction production is encouraged under this procurement system as the building contractor, being in charge of design, can reap the benefit of innovative products and processes.
- Late supply of information under this procurement system becomes a matter between the contractor and his or her building team.

Disadvantages

- The building contractor's in-house design expertise may be insufficient to solve the client's construction project needs efficiently.
- Tendering cost are high as contractors must design and produce accurate proposals as well as estimates.
- An inexperienced client still requires the expertise of professional advisers to prepare the briefing document, tender information and to evaluate quality and cost of design.

- The architect's professional indemnity insurance cover is assumed by all who have a design input in the proposed construction project, and this can be a huge burden on the small sub-contractor who is a party to the design.
- The building contractor requires an adequate insurance to cover design failures as he or she assumes the role of design as well as construction.
- Tender comparison becomes complex as it involves evaluation of design, quality and construction cost.
- Responsibility for defective design can be complicated by liability dates and time limitations.
- The client may become stranded with a construction product which is unsuitable for his or her needs.
- The client will find it difficult and/or costly to introduce variations once production has commenced on site.

15.3 DEVELOP AND CONSTRUCT (THE BPF SYSTEM)

The British Property Federation (BPF) have set out in their manual *System for Building Design and Construction* a construction procurement procedure aimed at producing construction projects quickly and economically and, hence, giving a better deal for construction clients. The manual proposes the appointment of a *client's representative* who is to be responsible for managing the construction project on behalf of the client. The manual further suggests the selection of a *design leader* who is to take the overall charge of the pre-tender design. Under the BPF system, the detailed design of the project is completed by the successful contractor and sanctioned by the design leader. The manual also sets out various processes and responsibilities aimed mainly at the alleviation of duplication of efforts by project participants, and these proposals may be summarised as follows:

- The design and production process is divided into five stages of conception, preparation of brief, development of design, tender documentation and site production.
- The responsibilities of all the project participants are defined and they are rewarded according to their performance.
- The duties of each party at each stage of the development process is clearly specified and, the working relationship between client, consultants and contractor is also defined.
- The architect's responsibility is limited to pre-tender design and sanctioning the contractor's detailed drawings. Moreover, architects are required to produce complete production information before the issue of tender documentation. This proposal is meant to encourage good decisions at the onset of the construction project and, hence, avert late expensive variations.
- Priced schedules of activities are employed in lieu of conventional bills of quantities.

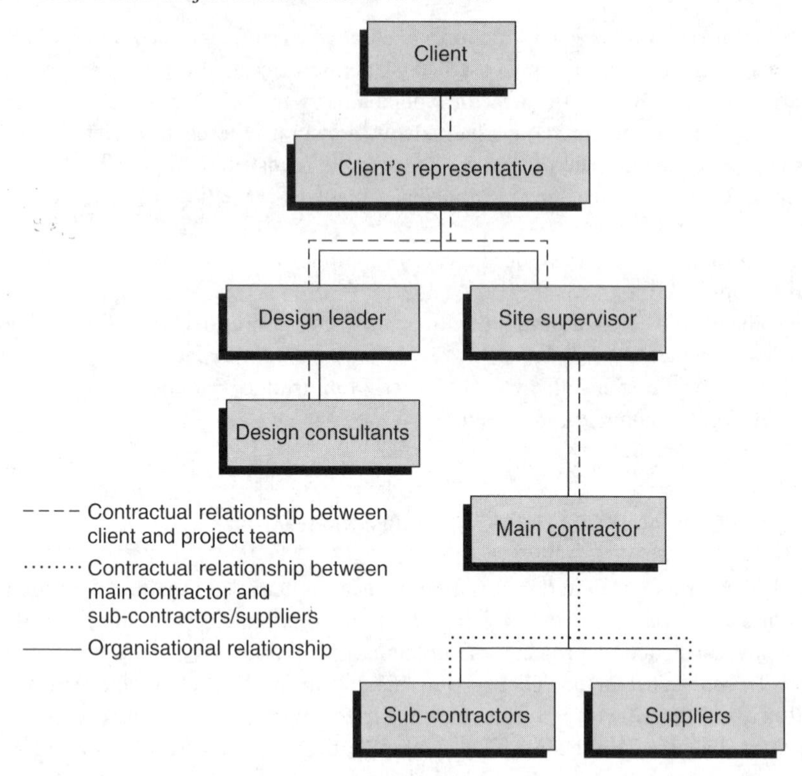

Fig. 15.3 Organisation structure of the BPF system

- Prime cost and provisional sums are eliminated and, hence, the contractor assumes full responsibility for selecting the specialist sub-contractors.
- Tenders are invited on complete production information and, in order to avoid any claims for extensions of time owing to lack of production information, the successful building contractor is required to sign a document stating that the drawn information is adequate for the execution of the works.
- The management of the project is the responsibility of project staff in the client's establishment, but where the client has no skilled staff in-house, an outside firm of consultants with skill and integrity is selected to perform the project management role.

15.3.1 Develop and construct (the BPF system) in practice

Like all construction projects, the client is the initiator of the construction process under the BPF system. However, according to the BPF manual, after the client has conceived the development, he or she develops or engages others to develop the concept and assess the technical feasibility and financial viability of the construction project. Moreover, at the early stage of the construction project, financial appraisal is normally undertaken followed by the preparation of an outline cost plan. The cost plan is updated as the project develops and cost information becomes available. When the viability of the proposed development project has been established, the client appoints a representative who may be one of his or her own staff, a specialist consultant or a firm. The function of the client's representative is to define the client's needs, resolve conflicting priorities and, at later stages, to instruct the specialist consultants and contractor. In all, the client's representative has emerged as the construction professional who exercises the role of a project manager plus many of the functions currently exercised by the architect and the quantity surveyor. For this reason, he or she is authorised to give instructions and deal with contractual claims and disputes of all kinds during the progress of the works. The criteria for selecting the client's representative are diverse but may include experience, technical competence, management ability and integrity.

Pre-contract phase

Briefing stage At the briefing stage, the client briefs the representative of the project's requirements. The client's representative, assisted by a design leader (a person or a firm with responsibilities for the pre-contract design and contract documentation functions) develops the brief, which should include a detailed statement of the client's requirements in terms of function, cost and time of the project. The briefing should be adequate to enable the design leader to develop the architectural design within the client's cost limit and obtain outline planning consent. However, as the prospective building contractor will be responsible for the detail design, the client's representative decides roughly the extent of design work to be assigned to the design team leader. In addition, the client's representative prepares the following documents:

1. *Master programme:* This contains a schedule of main items of work required in the design and production of the building.
2. *Master cost plan:* This contains an itemised forecast expenditure on design and production.

It is also the responsibility of the client's representative to recommend the appointment of the design leader (an individual or a firm) who takes on the total contractual responsibility for the pre-tender architectural and engineering design for a fixed fee. However, at times the client may prefer to have separate contracts with each of the specialist consultants engaged on the construction project. Should the latter be the client's preferred option for the appointment of consultants, the representative normally appoints one of the consultants to take the responsibilities of design leader.

Under the design leader's direction, the specialist consultants produce all scheme design drawings, obtain detailed planning consent, prepare cash flow forecast, cost monitoring and prepare contract documentation for obtaining tenders. As under the system the contractor is obliged to design some sections or details of the works, the contract documentation should clearly define those elements to be designed by the specialist consultants. The design of any element not so described becomes the responsibility of the contractor. In other words, the successful contractor is obliged to complete the design work, so far as is necessary, to enable him or her to complete the construction of the project.

Contractor selection Contractors are selected through single-stage selective tendering using drawings and specifications prepared by the client's specialist consultants under the direction of the design leader. The necessary documentation by which competitive tenders are obtained comprises drawings, specifications, conditions of contract, letter of invitation.

The tender After the preliminary enquiry to suitable building contractors, between three to six construction companies are short-listed and invited to tender for the construction project. As traditional bills of quantities are not in use under the BPF system, each tenderer is required to produce a priced schedule of activities with the tender submission. For this reason, each tenderer's quantity surveyor/estimator prepares, measures, quantifies and prices a schedule of activities. The schedule of activities comprises a schedule of the design, management and production activities that the tenderer has to carry out to accomplish the project. These, in a way, fulfil the role of priced traditional bills of quantities. During the preparation of the schedule of activities, senior managers, planners, and the plant and buying departments assist in the assessment of the extent of design, planning and work activities. Examples of some of the activities found in a schedule of activities are as follows:

- Site set up.
- Construction of foundations.
- Construction of superstructure.
- Supplying and fitting windows.

As the schedule of activities is intended, in part, to reflect the completed stages for which payment may be claimed, each activity should provide information regarding quantities, starting time, duration, resources, method statement, timing, sequence and price. Moreover, as the schedule of activities is one of the most important factors considered in the evaluation of tenders, the contractor may also use it directly for monitoring/controlling costs and progress during production. For this reason, the adequacy of information contained in a schedule of activities enhances the bidding contractor's chances of winning the project. Also, an adequately prepared schedule of activities facilitates the post-contract administration of construction projects. The sum of all the schedule of activity prices, plus a building contractor's profit and overheads, should be (or is equal to) a tenderer's total lump sum tender price.

Tender evaluation The evaluation of the tender is carried out by the design leader with the assistance of the other specialist consultants. The design leader then submits his or her findings and recommendations to the client's representative for approval/acceptance.

Production phase

Once production starts on site, the client's representative is responsible for the administration of the project. Normally, a project supervisor is appointed to the site to ensure that the project is executed in accordance with the contract specification.

Under the BPF system, the role of all consultants ends at the commencement of production. However, upon the advice of the client's representative, the client may decide to retain some of them to provide a post-contract service on the construction project. For example, the design leader may be retained to sanction the contractor's detail design and, at times, the quantity surveyor may also be retained to assist in the valuation of variations, preparation of interim valuations and final account, and the provision of general cost advice. Nonetheless, the client's representative sanctions all variations, approves all payment, settles the final account and carries out the final inspection for handing over.

Advantages and disadvantages

The advantages and disadvantages of the BPF system may be summarised as follows:

Advantages

- The contractor's expertise in detailing, quality and buildability is utilised as he or she contributes effectively to the detailed design of the construction project.
- The contractor's detailed design input leads to a shorter development time.
- The contractor is offered the financial incentive and encouragement to adopt the economies in design detailing and buildability.
- The contractor's design input enables contentious design solutions to be negotiated before contract.
- The contractors competition on price, as well as on adequacy of schedule of activities, resource allocation and method statement, enables the evaluation of the contractor's production and management skills.
- The contractor's responsibility for the selection of sub-contractors enables him or her to work with firms of known ability.

Disadvantages

- A high tender cost is incurred as design and measurement costs form part and parcel of the tenders submitted.
- The contractor's design input complicates the apportionment of design liability.
- The design responsibilities forced on the contractor requires his or her early involvement at the design phase, but this early involvement does not always occur.
- The contractor receives no payment until an activity is completed and this affects the contractor's cash flow on stretched activities.
- While the contractor may be competent as a constructor, he or she may not be equally as competent when it comes to dealing with conceptual design solutions.
- The contractor's choice of materials, design input and other recommendations may be biased towards manufacturers who offer him/her high trade discounts.

15.4 FAST TRACK SYSTEM

The fast track system is a managerial approach to the achievement of early construction project delivery. This system involves the application of innovation in the management of construction procurement and advances in the industrialisation of the construction process and, as a result, brings the following into play:

- The integration of design and production phases of construction projects.
- The involvement of the building contractor in both the design and production phases.
- Work packaging – the arrangement which breaks down works into trades or skills, or to a group of closely related trades or skills.
- Overlapping the work packages to enable production of sections of the construction project to proceed while the design for other sections is being considered or progressed (see Figure 15.4(c)).
- The employment of the expertise of works contractors and the recognition of their active participation in both design and production phases of construction projects.

Fast tracking is not a method of producing buildings cheaply; rather it is a system designed to increase the rate at which a construction project can be built and, hence, it costs more to adopt. The reason for its high cost is that the higher the rate of production rises above the optimum level of output – i.e. the level at which marginal increases in productivity become disproportionately expensive. Instead of maximising the output per worker, the building contractor puts the emphasis on increased total output in a given period. For this reason, more labour power is used, the working day is prolonged (by overtime or shift working), productivity suffers and the end result is a nominal increase in total output.

Generally, the fast track technique costs more to adopt because the expedited production is above the optimum level of production; however, the method's advantage lies in its reduction of development time. Therefore, if a construction client feels that speed is essential, it is necessary for his or her professional advisers to carry out an approximate cost–benefit analysis. The analysis should be based on the value of early possession and/or letting, leasing or sale as compared to the higher costs, risks and loss of financial control associated with fast tracking. If the benefits to be accrued from the former is more than the cost of the latter, the adoption of the fast track technique will be advantageous to the client. When the fast track technique is adopted, it is normally deemed essential to rationalise the management structure and centralise responsibility and decision making. For this reason, the client places the responsibilities of site identification, land assembly, appraisal, funding, design, construction and marketing on an individual (the project manager). He or she is also charged with the task of helping to establish the overall project objectives and of directing,

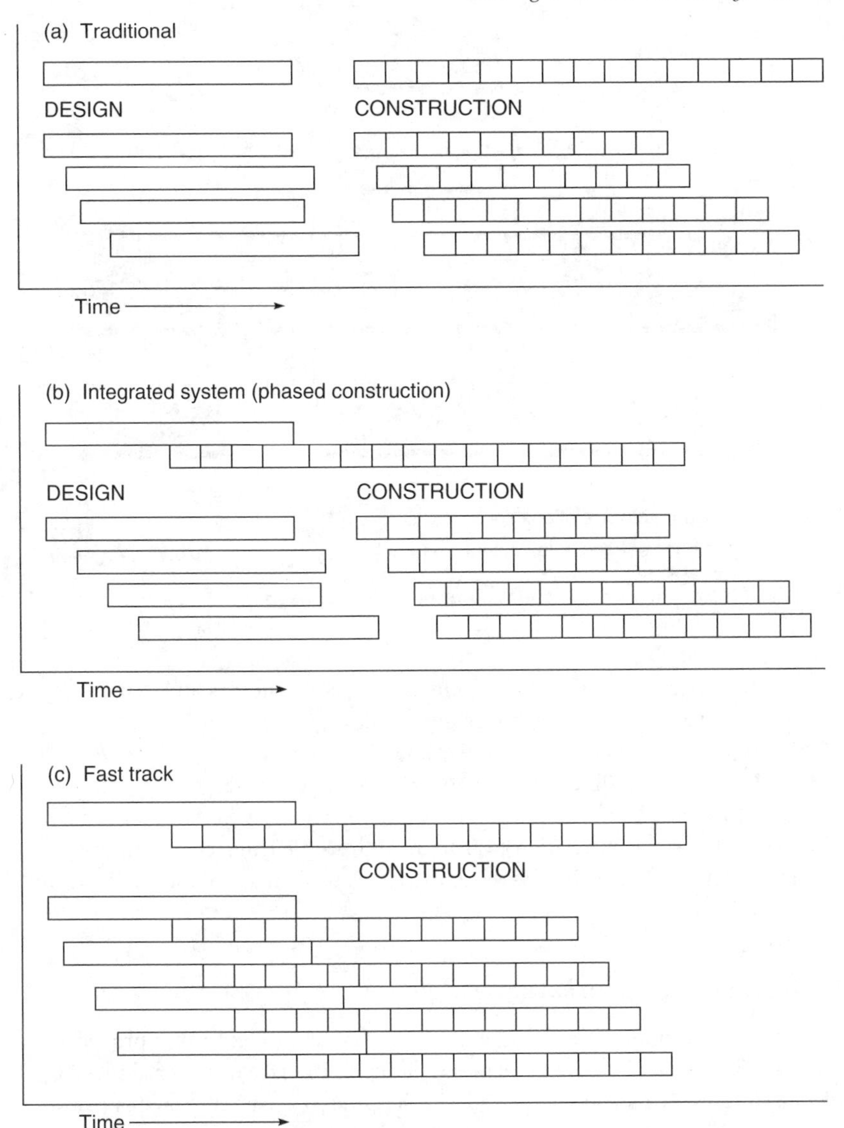

Fig. 15.4 (a) Traditional, (b) integrated and (c) fast track approaches compared

controlling, planning and coordinating the efforts of all participants to the achievement of those objectives. The fast track technique, therefore, normally employs the project management approach as illustrated by Figure 15.5, but for an accelerated construction programme. However, some fast track projects have, nevertheless, been executed adopting traditional management approaches.

In order to perform the duties efficiently, the project manger in a fast track construction project should have the authority to:

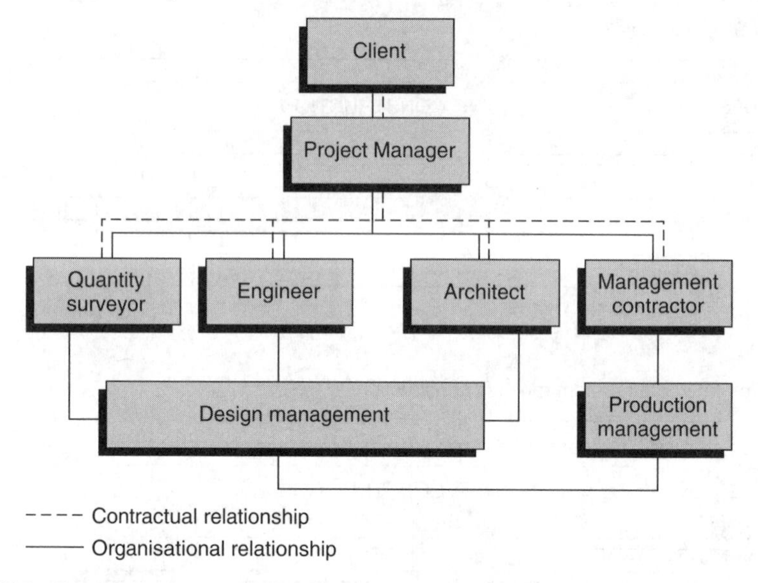

Fig. 15.5 Organisation structure for executive project manager

- Use funds, personnel and other resources subject to budgets and plans approved by the construction client (or his or her employer).
- Lead project personnel and set up work targets for them.
- Make decisions regarding variations to construction projects within limits approved by the client.
- Represent the client in relation to matters and people connected with the construction project, such as with ministries, consultants, building contractors and suppliers.

15.4.1 The emergence of fast tracking

The fast track technique was originated in North America and it has proved its ability to boost a client's development ambitions in the face of uncertainties of inflation, interest costs and competition in the property market. For this reason, it has been employed on major construction projects in America, Europe and other parts of the world with remarkable success. The main factors which influence the use of the fast track technique are:

- The need for early construction project delivery.
- Technical complexity of the construction project.
- The need for early appointment of the building contractor to contribute to the design of the construction project.
- Client possession of a flexible budget in addition to understanding the system.
- Where the construction project is prestigious and the image is crucial (e.g. head offices of multi-national companies).

15.4.2 Fast tracking in practice

After the client has established the need for a construction product, a project manager is appointed and the client briefs him or her of the project's requirements. The project manager helps to formulate the client's brief, advises on the appointment of the design team (architect/surveyor and engineers) and arranges for the following development activities.

Pre-contract phase

Design As the project leadership is assumed by the project manager under the fast track approach, the architect's role in the fast track project is confined to designing, writing specifications and obtaining planning consents. At the construction project's design phase, the quantity surveyor prepares the preliminary budget for the scheme and, where time permits, may prepare an outline cost plan.

Selection of management contractor As discussed in the management contract construction procurement system, the management contractor is selected through a two-stage selective tendering option. This procedure of selection enables the project manager to assess the management contractor's ability to make technical and managerial contributions to the construction project design and production in addition to working within a tight budget. On appointment the management contractor joins the design team and provides advice on a buildable design and later manages the various works contractors.

Selection of works contractors Works contractors compete for and execute the individual work packages. However, so different from traditional practice are the qualities demanded of works contractors by fast tracking that new selection criteria are appropriate. For this reason, price is not the only determining factor when selecting works contractors. Indeed, so crucial are other criteria that competitive tendering may be waived in favour of intensive negotiation and interviewing. In particular, prospective works contractors should demonstrate their ability to:

- Provide the technical and managerial capabilities required for delivery on a tight programme.
- Understand the design implications and coding with particular reference to interfacing.
- Offer the requisite design input.

Production phase

At the production phase, periodic site inspections are carried out by the architect and the project manager to ensure that the quality specified in the contract documentation is being met and that good construction methods are being practised.

The quantity surveyor prepares periodic valuations for interim payments, values all authorised variations in the project and prepares the final account. It is also his or her responsibility to evaluate all contractual claims.

The structural and services engineers conduct periodic checks and tests on-site to ensure compliance with contract specification. In addition they provide professional advice on matters relating to their respective fields of expertise.

Commissioning stage

At the commissioning stage, the architect and the project manager ensure that the client obtains all relevant documents such as as-built drawings and operation/ maintenance manuals.

Strategic options In contrast with the project management system, the fast track technique possesses several strategic options. For a typical fast track project the desired objective may be achieved by adopting the following:

- Simplification of the brief; however, it should be adequate for development.
- Early appointment of the managing contractor to contribute to design and advise on availability of resources.
- Pre-purchasing essential construction materials and components when required.
- Selection of works contractors based on the ability to work within tight budget and make technical and managerial contributions to the project design and production.
- Preference for steel frame to reinforced in-situ concrete as steel is faster to erect and supports itself and other components easily.
- More use of prefabricated materials and components and avoidance of wet trades.
- Prompt decision making and flexible budgets should be made available for quick implementation of decisions.
- Odd hours (night and weekends) as well as shift working planned for certain operations.
- Shell and core approach adopted to enable prospective tenants to choose the equipment and partitioning which benefit their line of business.
- The higher than average intensity of work planned for requires continuous and efficient utilisation of resources.
- Variations at construction phase are kept to a minimum in order to maintain the tempo of production.

The above courses of action are taken to ensure that the construction project is completed on or before time to enable the client to benefit from the cost of expedited construction. Moreover, during the progress of the works, the project manager carries out his or her contractual responsibility of supervising the project and its participants and, driving it to a successful completion.

Advantages and disadvantages

Advantages and disadvantages of the fast tracking technique may be summarised as follows:

Advantages

- Early completion of construction project enables the client to put his or her building on the market before the buildings of competitors.
- By compressing the construction process, the client can gain early occupation of the building.
- In the case of buildings for sale or rent, return on capital employed is quicker than under traditional systems.
- A compressed production programme may beat the deadline for statutory or market changes affecting new developments (e.g. increases taxation, interest rates and resource prices generally).
- Early completion enables the client to meet current demands for development (i.e. before competitors can expand their production lines).
- An expedited production programme enables the client to reduce uncertainties of development quickly.
- Early completion leads to savings in the time-related preliminary items.
- The client saves interest payment on finance for land purchase.

Disadvantages

- Production starts on partial design and this involves the client in the high risk of not knowing precisely what is to be built or at what cost until production is far advanced.
- The client pays extra finance charges on funds for up-front expenditure on advance purchasing of essential materials/components and accelerated cash flow.
- It is an expensive method to adopt as production is above the optimum level of output – i.e. the level at which a marginal increase in productivity becomes disproportionately expensive.
- The client may be subjected to higher costs, risks and loss of financial control.
- Design becomes fragmented as works contractors are required to accept design responsibility for which they are not well equipped, and liabilities for defective design may be difficult to ascertain.
- The client may be uncertain as to his or her financial commitment until a late stage.
- The management contractor avoids most of the contractual risks borne by the main contractor under the traditional system and this places the client in a precarious position.
- The client pays additional fees to the project manager and the management contractor; he or she also pays double insurance for design (one for consultant design work and the other for works contractors' design work) and double preliminaries for management contractor and works contractors.

SUMMARY

In this chapter we have discussed some of the other non-traditional methods of construction procurement currently being used along with the traditional methods. Like the integrated construction procurement system, the procurement methods discussed above encourage an active participation of the building contractor at both design and production phases of the construction process. Hence, under these non-traditional systems, the building contractor is involved in some aspects of design which eventually leads to production economies. The other aspect of these non-traditional systems is the overlapping of the design and production functions which eventually lead to a shorter project delivery time. However, this approach is not without the danger of the client and building contractor agreeing to commence production on a half design or on uncompleted project information.

FURTHER READING

Davies, C. 1984 The BPF system: the system in action. *Building*, June, pp. 31–3.
Davies, C. 1984 The BPF system: no going back. *Building*, June, pp. 28–9.
Franks, J. 1992 Design and build tendering – do we need a code of practice? *Chartered Builder*, June, pp. 8–10.
Griffith, A. 1989 Design–build: procurement and buildability. *CIOB Technical Information Service, No. 112.*
Hughes, W.P. 1992 An analysis of construction management contracts. *CIOB Technical Information Service, No. 135.*
Kwakye, A.A . 1991 Fast track construction. *CIOB Occasional Papers, No. 46.*
Minogue, A. 1992 Don't just blame design and build. *Building,* June, pp. 32–3.
Pain, J. 1993 Design and build compared with traditional contracts. *The Architects Journal,* November, pp. 34–5.
Pennington, I. 1983 *A guide to the BPF system and contract*. CIOB.
Titmus, P.D. 1982 Design and build in practice. *Building Technology and Management,* April, pp. 9–12.
Waters, B. 1983 The BPF system. *Building,* December, pp. 25–30.

CHAPTER 16

EUROPEAN CONSTRUCTION PROCUREMENT

16.1 INTRODUCTION

In the foregoing chapters we have discussed some of the procedures and techniques normally adopted in the procurement of construction projects in the UK. This chapter has been written with the intention of giving readers a basic awareness of the construction industry and construction procurement procedures in other member states of the European Union (EU). However, it must be mentioned at this point that, like the UK, the construction industries and procurement practices in the other member states of the EU are complex and, therefore, it is not possible to deal with all the intricacies within the confines of this book. The exposition does, however, enable the reader to draw some comparisons with the construction procurement practices and administration operating in the UK as discussed in the foregoing chapters.

16.2 THE INFLUENCE OF THE EU ON THE CONSTRUCTION INDUSTRY

The European Union has yet to establish a common construction industry. However, Article 59 of the Treaty of Rome prohibits the prevention of free movement of goods between member states. Article 7 of the same treaty prescribes that people must be free to offer their goods or services in other member states without any form of discrimination or restrictions. The above EU requirements, therefore, have a major impact on the construction industry and their directives dealing with public procurement of goods, services and construction products are meant to achieve the aims of free trade between member states.

16.2.1 The public procurement directives

The public procurement directives are intended to achieve the objectives of the Treaty of Rome. Hence, it opens up public procurement within member states

Table 16.1 The EU public procurement directive relevant to the construction industry

Directive title	Directive purpose
Supplies directive and amendments	Deals with public supplies/products for the public sector.
Services directive	Concerned with the procurement of services by public bodies, including maintenance and repair.
Utilities directive	Concerned with the procurement of supplies and works contracts and also, private purchases in the water, energy, transportation and telecommunications sectors.
Utilities remedies directive	Enables aggrieved parties to seek redress when they consider that this directive has been infringed.
Works directive and amendments	Deals with public works contracts awarded by state and regional or local authorities.

for greater competition from non-national companies. The EU public procurement directives which are relevant to the construction industry are shown in Table 16.1.

16.3 TENDERING ARRANGEMENTS

The EU directive on public procurement provides for open competitive tendering for all construction projects of a certain monetary value undertaken by public bodies within the member states of the EU. However, the public works directive and amendments allows the use of restricted and negotiated tenders only in justified circumstances. It obliges the member states to adopt their tendering procedures for public works contracts to all the rules laid down in the directive in order to make that awarding process accessible and transparent to all potential bidders throughout the Union. This directive even applies to contracts awarded by private persons but which are financed (for more than 50 per cent) by any of the above-mentioned public authorities when these contracts concern educational and administrative buildings and also, sports and recreation facilities. For this reason, the Works Directive and Amendments provides for the advertisement of all contract works falling within its scope without charge for the official journal of the European Union. The advertisement should contain adequate information on the technical specification, the proposed procurement programme, the tendering procedure to be used, the tendering period rules on participating and criteria for award. The criteria for the award of a contract must be either the lowest price only, or various criteria which must be determined in advance in decreasing order of importance when the award is for the economically most advantageous tender.

Additionally, the transparency of tendering and award procedures is guaranteed by the obligation of contracting authorities:

- To explain at an eliminated contractor's expense why his or her bid has been rejected.
- To make a report on each award decision and supply it to the Commission on request.
- As mentioned above, to publish a notice on the outcome of each award decision.

The above directives and their numerous procedures are not intended to create a common public procurement policy for the member states. Rather, they are a means of establishing a basic set of common rules in clear understandable terms for public sector projects which fall within the scope of the EU directives. The directives also ensure that every company in the member states is given a fair chance to compete for projects in the EU without any form of discrimination or restriction on grounds of nationality.

16.4 METHODS OF PROCUREMENT

The European Union has yet to establish a common construction procurement method; however, the component tasks of the construction procurement which comprise inception, design, documentation, production and commissioning are common to most construction projects and are undertaken in all member states. The only difference is that these tasks are undertaken under diverse economic and physical climates, to different standards and regulations and also under different contractual and organisational arrangements. Furthermore, these similarities in the component tasks encourage competition and also facilitate the execution of construction projects by non-national construction companies across the common borders of member states of the EU. A summary of construction procurement procedures adopted in the UK and the other member states of the European Union is given in Table 16.2.

Table 16.2 Summary of UK and EU construction procurement procedures

Country	Type of contract					Tendering method		
	Lump sum	Trade system	Measurement	Cost reimbursement	Package deal	Open	Selective	Negotiated
Belgium	✓		✓			✓	✓	✓
Denmark	✓	✓		✓	✓	✓	✓	✓
France	✓				✓	✓	✓	✓
Germany	✓		✓		✓	✓	✓	✓
Greece	✓					✓	✓	✓
Ireland	✓		✓	✓	✓	✓	✓	✓
Italy	✓				✓	✓	✓	✓
Luxembourg		✓				✓		✓
The Netherlands	✓					✓	✓	✓
Portugal	✓	✓				✓	✓	✓
Spain	✓					✓	✓	✓
United Kingdom	✓		✓	✓	✓	✓	✓	✓

16.5 CONSTRUCTION PRACTICES

As noted above, although the component tasks of construction procurement are common to all construction practices of member states of the EU, they are organised under different standards, regulations and liabilities. The following discussion is therefore aimed at providing an insight into the key features of the construction industry and procurement practices within member countries of the EU.

16.5.1 Belgium

Structure of the construction industry

The Belgian construction industry is made up of a mixture of large and small construction companies. However, a few of the larger construction companies make up the backbone of the industry. The Belgian construction industry also employs 5.6 per cent of the nation's workforce. In addition, construction professionals (architects and engineers), whose roles are defined and regulated by law, play an important function in the activities of the Belgian construction industry.

Parties to the construction process In Belgium, the parties to the construction process comprise the client, the architect and the building contractor.

Design

The design of buildings is carried out by a university trained registered architect who signs drawings. The architect is obliged to defend the public interest as well as that of the client and, for this reason, gives a ten-year guarantee for the stability of his or her design. However, on major building projects, an engineer also participates in the design function.

Types of contract

Contracts in regular use are:

- A strictly fixed lump sum contract *(Marche à Forfait Absolu)* with no provision for variations in quantities or contract sum.
- A lump sum contract *(Marche à Forfait Relatif)* with provisions for both variations and fluctuations.
- The remeasurement contract *(Marche à Borderaux de Prix)* is based upon a schedule of rates.

Each of the above contracts has a standard form of contract which defines the tasks, obligations and responsibilities for the client, architect and the building contractor.

Methods of obtaining tenders

Methods of tendering adopted include:

- Selective tendering – used for private clients' projects.
- Negotiated tender – used for private clients' projects.
- Public tendering – used for government projects.

Tender documents The tender documents are prepared by an architect or an engineer and are composed of drawings, form of contract, general specification for materials and workmanship and a particular specification.

Tender action The bidding contractor submits a lump sum price for the execution of the works and, in a lump sum contract, he or she supports his or her tender with a schedule of quantities and rates which, when successful, will form the basis of the valuation of variations.

Production and payment

During the progress of the works, the architect provides general direction in lieu of permanent supervision. The contractor supervises and executes the works and, in return, receives periodic interim net payments (i.e. payment less 10 per cent retention) based on stages of work completed. The client releases the sum retained to the building contractor on satisfactory completion of the works.

16.5.2 Denmark

Structure of the construction industry

The Danish construction industry is composed of a mixture of large and small construction companies and employs approximately 7 per cent of the nation's workforce. In addition, construction professionals (architects and engineers), whose roles are defined and regulated by law, play an important function in the activities of the industry.

Parties to the construction process In Denmark, the parties to the construction process comprise the client, architect and the building contractor.

Design

The design of buildings is carried out by a university trained, registered architect. However, an engineer's input is required in the design of major building projects.

Types of contract

Contracts in regular use are:

- Cost reimbursement contract.
- Lump sum contract.
- Trade contract system.
- Design and build.

The general conditions for the provision of works and supplies within building and engineering are covered by Contract AB92 issued by the Danish Ministry of Housing.

Methods of obtaining tenders

Methods of tendering adopted include:

- Open tendering – used for most government projects.
- Selective tendering – used for government projects.
- Negotiated tender – used for private clients' projects.
- Package deal tender – used for private clients' projects.

Tender documents The tender documents are prepared by an architect or an engineer and are composed of drawings, detailed specification, form of agreement, form of tender and general conditions of contract.

Tender action The bidding contractor prepares his or her own schedule of quantities from which a lump sum price is submitted for the execution of the works. In addition, the successful bidder is required to provide a performance bond of 15 per cent of the contract sum.

Production and payment

During the production phase, the architect undertakes the technical control, planning and supervision of the construction work. The building contractor, on the other hand, executes the works and submits periodic interim applications for payment of 90 per cent the work satisfactorily completed. The architect or the engineer checks all the contractor's interim applications before payment by the client. The money retained is released to the contractor on satisfactory completion of the works.

16.5.3 France

Structure of the construction industry

The French construction industry is made up of a mixture of large and small construction companies. However, there is less use of sub-contracting and, hence, many projects are dealt with through trade contractors; the coordinator for this arrangement is either the architect or the engineer. About two million of the working population of France are employed in the construction industry. In addition, construction professionals (architects and engineers), whose roles are designed and regulated by law, play an important function in the activities of the French construction industry.

Parties to the construction process Parties to the construction process in France consist of the client, the architect and either the main building contractor or separate trade contractors.

Design

The design of buildings is carried out by a registered or educated architect who is obliged to insure the design for ten years against the stability of his or her design. Engineers also participate in the design of major building projects.

Types of contract

Contracts in regular use are:

- Lump sum contract (with provisions for variations and fluctuations).
- Package deal contract.

The main standard form of contract is the *Code des Marches Publiques* on which private sector contracts are based.

Methods of obtaining tenders

Methods of tendering adopted include:

- Open tendering *(L'adjudication)* – this is used for government projects.
- Selective tendering *(L'adjudication)* – this is also used for government projects.
- Negotiated tender – used for private clients' projects.

Tender documents The tender documents are prepared under the direction of either an architect or an engineer and are composed of drawings, specification, national codes of practice, contract conditions, contract agreement and form of tender.

Tender action The bidding contractor prepares his or her own schedule of quantities and submits a lump sum price for the execution of the works. The bidding contractor is also obliged to support the tender with a detailed breakdown of the tender as well as supplying rates for valuing variations should he or she be successful.

Production and payment

During the production phase, the architect is responsible for ensuring that the work proceeds as planned. While on site, the building contractor is usually responsible for the detailed design and also the preparation of interim applications for payment of 90 per cent of the works satisfactorily completed. The interim application is checked by the engineer or the architect before payment is made by the client. The money retained is released to the contractor on satisfactory completion of the works.

16.5.4 Germany

Structure of the construction industry

The German construction industry is made up of few large and several small construction companies. About two million of Germany's working population is employed in the construction industry. In addition, construction professionals (architects and engineers), whose roles are defined and regulated by law, play an important function in the activities of the German construction industry.

Parties to the construction process In Germany, the parties to the construction process comprise the client, the architect and either the main building contractor or separate trade contractors.

Design

The design of buildings is carried out by a registered architect who is required to take out a professional indemnity policy. However, this indemnity period may be reduced by agreement with the client. The architect also prepares the budget and provides cost advice at the project's feasibility stage.

Types of contract

Contracts in regular use are:

- Measurement contract.
- Fixed lump sum contract (which allows rates alteration to reflect variation of quantities in excess of 10 per cent).
- Package deal contract.

Method of obtaining tenders

Tendering procedure is laid down in *Verdingungsordung für Banleistungen (VOB)* and the method of tendering adopted includes:

- Open tendering – used for government projects.
- Selective tendering – used for private clients' projects.
- Negotiated tender – used for private clients' projects.

Tender documents The tender documents are prepared by an architect or an engineer and they comprise the VOB (which is in three parts: (1) general conditions; (2) general procedures; (3) general technical regulations), drawings and an approximate schedule of quantities.

Tender action During the tender period, the bidding contractor prices an approximate schedule of quantities to submit a lump sum tender for the execution of the works.

Production and payment

At the production phase, the architect supervises the construction work and provides the client with periodic financial statements and cost advice. An interim payment equivalent to 95 per cent of work satisfactorily completed is paid to the building contractor as production progresses. The sum retained is released to the contractor on satisfactory completion of the works.

16.5.5 Greece

Structure of the construction industry

The Greek construction industry is made up of a mixture of medium and small construction companies and employs 6 per cent of the total Greek workforce.

Parties to the construction process In Greece, the parties to the construction process comprise the client, the architect and the building contractor.

Design

The design of buildings is carried out by a qualified architect. However, an engineer's input is required in the design of major building projects.

Type of contract

The contract in regular use is the lump sum contract (with provision for both variations and fluctuations).

Methods of obtaining tenders

Methods of tendering adopted include:

- Open tendering – used for government projects.
- Selective tendering – used for both private clients' and government projects.
- Negotiated tender – used for private clients' projects.

Tender documents The tender documents are prepared by an architect or an engineer, and usually comprise drawings, general specification, technical specification, technical description, work item description and unit price list, schedule of quantities and cost estimate, form of tender.

Tender action The contractor prices a schedule of quantities to arrive at a lump sum price for undertaking the construction project.

Production and payment

During the production phase, the architect supervises the smooth running of the project. While the work is in progress, the building contractor receives a periodic interim payment prepared by an engineer. This payment is equivalent to 90 per cent of work satisfactorily completed and the retained amount is released to the contractor on satisfactory completion of the works.

16.5.6 Ireland

Structure of the construction industry

The Irish construction industry is made up of a mixture of large and small construction companies. This industry employs 6.5 per cent of the nation's workforce.

Generally

The structure of the construction industry, the organisation of the construction process, the professions and even the contractual procedures adopted, closely resemble those in use in the UK construction industry.

16.5.7 Italy

Structure of the construction industry

The Italian construction industry is highly fragmented, comprising a handful of large groups and thousands of small to medium sized construction companies. Employment in the Italian construction industry is equal to 9.4 per cent of the nation's workforce.

Parties to the construction process In Italy, the parties to the construction process comprise the client, the architect and the building contractor.

Design

The design of buildings is carried out by a graduate architect who signs the drawings and has ten years' liability for his or her work. However, on major building projects, an engineer participates in the design.

Type of contract

Contracts in general use are:

- Fixed lump sum contract (at times with fluctuation provisions).
- Package deal contract.

Methods of obtaining tenders

Methods of tendering adopted include:

- Open tendering – used for government projects.
- Selective tendering – used for government projects.
- Negotiated tender – used for private clients' projects.

Tender documents The tender documents are prepared by an architect or an engineer and comprise drawings, general form of contract conditions, special specification clauses (applicable to the particular project) and form of tender.

Tender action The bidding contractor prepares his or her own schedule of quantities from which a lump sum price is submitted for the execution of the works.

Production

It is the architect's responsibility to supervise the works during production. As the work progresses, periodic interim payments equivalent to 90 per cent of the completed work are paid to the building contractor. The sum retained is released by the client on satisfactory completion of the works.

16.5.8 Luxembourg

Structure of the construction industry

Being a smaller economy, Luxembourg's construction industry is generally of limited size and capacity. Although there are a limited number of construction companies of a mixture of small to medium size, it plays host to many foreign construction companies and consulting firms.

Parties to the construction process In Luxembourg, the parties to the construction process comprise the client, the architect, the building contractor and sub-contractors.

Design

The design of building is carried out by a consultant architect who (when required by the client) takes out a ten-year insurance cover against the stability of his or her design. On major projects, however, an engineer's input is also required in the design.

Type of contract

Generally, apart from major projects which are usually undertaken by main contractors, the trade contract system is regularly used.

Method of obtaining tenders

Methods of tendering adopted include:

- Open tendering – used for government projects.
- Negotiated tender – used for both private clients' and government projects.

Generally

In Luxembourg, construction activities are not regulated by national law. Therefore, there is no national standard for technical specifications. For this reason, construction activities are regulated by each of the 118 authorities who set out their own regulations which mainly cover planning and technical requirements. However, Title 3 of the Grand-Ducal Regulations contains a set of General Contract Conditions *Cahier des Charges Fixant les Clauses et Conditions Générales des Marchés* for use by local authorities. But there is no standard contract for projects for the private sector clients. Generally, German and French standards of workmanship are widely accepted and, on major building projects, clients sometimes require building contractors to take out *all risk site insurance cover*. However, defects liability insurance, which provides cover for two or ten years, is regularly required for government projects.

16.5.9 The Netherlands

The structure of the construction industry

The construction industry of the Netherlands comprises a mixture of large and small construction companies and employs 7.4 per cent of the nation's workforce.

Parties to the construction process In the Netherlands, the parties to the construction process comprise the client, the architect and the building contractor.

Design

The design of buildings is carried out by an architect who also provides cost advice at the project's feasibility stage. Nevertheless, an engineer's input is required in the design of a major building project.

Types of contracts

Contracts in general use are:

- Fixed lump sum contract.
- Lump sum contract (with provision for both variations and fluctuations).

Method of obtaining tenders

Methods of tendering adopted include:

- Open tendering – used for government projects.
- Selective tendering – used for both private clients' and government projects.
- Negotiated tenders – used for private clients' project.

Tender documents The tender documents are prepared by an architect or an engineer and normally include drawings, form of contract and a general specification for material and workmanship.

Tender action The bidding contractor prepares his or her own schedule of quantities from which a lump sum price is submitted for the execution of the project. The bidding contractor is also required to support his or her tender with a schedule of quantities and rates in operational order.

Production and payment

During the production phase, the architect supervises the works in progress; the contractor prepares and receives interim stage payments equivalent to 90 per cent of work satisfactorily completed. The sum retained is released by the client on satisfactory completion of the works. On completion, the building contractor is required to insure the building against defects liability for a period of ten years.

16.5.10 Portugal

Structure of the construction industry

The Portuguese construction industry is composed of a mixture of a number of small to medium construction companies. Portuguese construction professionals (architects and engineers), whose roles are defined and regulated by law, play an important function in the activities of the industry.

Parties to the construction process In Portugal, the parties to the construction process comprise the client, the architect, the engineer, project coordinator and sub-contractors.

Design

The design of buildings is currently carried out by a registered architect but, in the past, the engineer had played a dominant role in this field. However, the engineer is still responsible for designing the building to withstand earthquake loading.

Types of contracts

Contacts in general use are:

- Separate trade contracting system (for major projects).
- Fixed lump sum contract.

Method of obtaining tenders

Methods of tendering adopted include:

- Open tendering – used for government projects.
- Selective tendering – used for both private clients' and government projects.
- Negotiated tender – used for both private clients' and government projects.

Tender documents The tender document is prepared by an architect or an engineer and it is composed of either standard general legal conditions, specifications and drawings or standard general legal conditions, specifications, drawings and a schedule of quantities.

Tender action The bidding contractor prices a schedule of quantities. But where tenders are sought on drawing and specification, the contractor produces and prices his or her own schedule of quantities and then submits a lump sum price for the execution of the works.

Production and payment

The responsibility for the site supervision during production is mainly undertaken by engineers or technical engineers. In addition, at the client's request, engineers undertake financial and quality control aspects of the project. The contractor organises and executes the works and, in return, receives periodic interim payments as work progresses.

16.5.11 Spain

Structure of the construction industry

The Spanish construction industry is made up of a larger number of small local construction companies and a small number of large ones. Construction professionals (architects and engineers), whose roles are defined and regulated by law, enjoy high status and play an important function in the activities of the industry.

Parties to the construction process In Spain, the parties to the construction process comprise the client, the architect and the contractor.

Design

The design of buildings is carried out by a qualified architect who is obliged to insure his or her design for ten years against serious defects. Engineers also participate in the design of major building projects.

Types of contract

The contract in regular use is the lump sum contact.

Methods of obtaining tenders

Methods of tendering adopted include:

- Open tendering (auction or competition).
- Selective tendering – used for both private clients' and government projects.
- Negotiated tender – used for government projects.

Tender documents The tender documents are prepared under the direction of an architect or an engineer and consists of the following:

- Book of general administrative clauses (*Pliego de Clausulas Administrativas Generales*).
- Book of special administrative clauses (*Pliego de Clausulas Administrativas Particulares*).
- Book of general technical prescriptions (*Pliego de Prescripciones Technicas Generales*).
- Book of particular technical prescriptions (*Pliego de Prescripciones Technicas Particulares*).

Tender action As estimated quantities are provided in the tender documents, the bidding contractor prices the quantities and thereby produces a lump sum price for the execution of the works.

Production and payment

During the progress of the works, the architect provides technical direction on site. The contractor assumes the responsibility of completing the detail design and the works. In return, he or she receives periodic interim payments (after remeasurement) for works satisfactorily completed.

SUMMARY

In this chapter, we have discussed the efforts made by the European Union to promote competition and free trade among member states. Furthermore, we have discussed, in outline, the various methods adopted in the procurement of construction project among the member states.

The main difference between the construction practice of the UK and that of the other member states of the EU is that the quantity surveying profession does not exist in other member sates. Apart from France, which utilises the services of a measure (*Metreur*) and verifier (*Verificateur*), which come nearest to the quantity surveying service in the UK, the quantity surveying functions in all member states are performed by either an architect or an engineer as part of their professional services. Even so, the French regard the services performed by the measures and verifiers as technical services as opposed to professional services performed by the quantity surveyor in the UK.

Furthermore, major differences exist in the legal basis of construction as each member state within the EU procures its construction project under a construction contract bodied in the law of its land. Even where parties to a construction project may have similar titles (i.e. architects, engineer, contractor) they have different responsibilities in the construction process in each member state. To complicate the issue still further, these responsibilities vary with different sectors of the industry, (e.g. public, private, civil engineering or building works) or in different regions of the member states. However, the good thing about it is that each of the EU member countries has developed its own approach to the complex problems associated with the administration of the procurement of construction products to meet the needs of construction clients and/or its ultimate users within their respective borders.

FURTHER READING

Borrie, D. 1995 Procurement in France. *CIOB Construction Papers, No. 49.*

Chapman, N.F.S.; Grandjean, C. 1981 *The construction industry and the European community.* BSP Professional Books.

Chapman, K.; Karadelis, J. 1994 Building in the home of gods: an examination of the construction industry in Greece. *CIOB Construction Papers, No. 42.*

Cooke, B.; Walker, G. 1994 *European construction procedures and techniques.* Macmillan.

Diggings, L. 1991 *Competitive tendering and the European communities.* Association of Metropolitan Authorities.

Ferry, D. 1991 *The organisation of procurement in civil engineering project procedure in the EC.* Thomas Telford.

Hardy, T.J. 1992 Germany challenge for the estimator. *CIOB Construction Papers, No. 11.*

Houtle, V.V. 1991 *The impact of Europe upon the construction industry.* Lloyds of London Press.

Spons, H. and B. 1983 *Spons architects and builders.* Price Books.

REFERENCES

Ahenkorah, K. 1994 Reimbursement for changes in construction costs. *The Building Economist*, March, pp. 19-22.

Ahenkorah, K. 1993 Re-thinking the retention rule. *Chartered Quantity Surveyor*, November, pp. 8-9.

Alexander, D. 1994 The accidental designer. *New Builder*, February, pp. 14-15.

Anderson, W.E.; Bailey, P. 1981 Productivity and the design team. *Building Technology and Management*, December, pp. 3-6.

Appleby, R.C. 1977 *Modern business administration*. Pitman.

Ashworth, A. 1994 *Contractual procedures in the construction industry*. Longman.

Ashworth, A. 1988 *Cost studies of buildings*. Longman Scientific and Technical.

Austin, B. 1991 True value. *New Builder*, March, pp. 26-7.

Austin, S. 1992 Time on your side. *New Builder*, April, pp. 16.

Baird, R. 1990 A better deal for the Client. *Chartered Quantity Surveyor*, July, pp. 9 and 13.

Balchin, P.N.; Kieve, J.L. 1979 *Urban land economics*. Macmillan.

Ball, M. 1988 *Rebuilding construction – economic change in the British construction industry*. Routledge.

Barda, P. 1991 Risk prevention. *The Building Economist*, September, pp. 27-9.

Barnes, M. 1988 *Construction project management*. Vol. 6, Butterworth, pp. 69-79.

Barthurst, P.E.; Butler, D.A. 1973 *Building cost control techniques and economics*. Heinemann.

Bashil, I.; Stimson, M. 1990 Arbitration – pros and cons. *New Builder*, November, pp. 32.

BEC 1991 Causes of conflict. *Building*, November, pp. 111.

Bennett, J. 1985 *Construction management and the chartered quantity surveyor*. RICS.

Bennett, J.; Flanagan, R. 1983 Management options. *Building*, April, pp. 32-3.

Beresford, T.; Kellard, P. 1992 What's allowed? *New Builder*, February, pp. 22.

Bernstein, R.; Wood, D. 1993 *Handbook of arbitration practice*. Sweet & Maxwell.

Bevan, A. 1992 *Alternative dispute resolution*. Sweet & Maxwell.

Bingham, T. 1991 A plain collateral contract to overcome everyone's prejudice. *Building*, March, pp. 33.

Bingham, T. 1990 No losers when commerce triumphs over litigation. *Building*, September, pp. 48.

Bingham, T. 1990 Industry divided by Cavemen tactics of high retention. *Building*, September, pp. 45.

Birchall, D.; Newcombe, R. 1985 Developing the skills. *Chartered Quantity Surveyor*, July, pp. 472-3.

Birnberg, H. 1993 Roles of a project manager. *Architecture*, July, pp. 115-17.
Bishop, D. 1994 Safety planning. *New Builder*, June, p. 12.
Bishop, D. 1974 Productivity in the construction industry. *Building*, August, pp. 77-8.
Bishop, D. 1994 Impositions and opportunities – Health and Safety. *Chartered Surveyor Monthly,* February, pp. 36-7.
Bloomfield, J. 1993 Value added quality. *Chartered Builder*, April, pp. 12-13.
Blyth, A. 1993 The architect and adjudication. *The Architects Journal*, July, p. 32.
Bone, S.; Loring, J. 1994 Meeting new responsibilities. *The Architects Journal*, July, p. 23-5.
Bone, S.; Loring, J. 1994 How to be a planning supervisors. *The Architects Journal*, July, p. 34-6.
Booth, S. 1991/92 Better contract terms – better project results. *Civil Engineering Surveyor*, December/January, pp. 5-6.
Borrie, D. 1995 Procurement in France. *CIOB Construction Papers, No. 49.*
Bowyer, J. 1992 *Practical specification writing*. Hutchinson.
BRECSU 1993 Energy efficiency in new housing. *Good Practice Guide No. 79*, Building Research Establishment.
Brewer, G. 1994 The new PSA/1 standard form of contract. *Construction Law*, pp. 7-9.
Briggs, B. 1989 Managing health and safety in construction. *Chartered Quantity Surveyor*, pp. 24–5.
Browne, K. 1992 Health check: fibres in the lungs. *The Architects Journal*, February, pp. 45-8.
Bunton, L. 1986 Contractual disputes. *Chartered Quantity Surveyor*, October, pp. 24-6.
Bunton, L. 1984 Avoiding disputes. *Building*, March, p. 25.
Burberry, P. 1977 *Environment and services*. Batsford.
Burch, T. 1991 Alternative dispute resolution. *Building*, November, pp. 12-13.
Butcher, T. 1991 Disputes – watch out for settlements that won't settle down. *Building*, June, p. 29.
Calvert, R.L. 1976 *Introduction to building management.* Newness–Butterworth.
Carter, M.R. 1993 *Resource management for construction*. Macmillan.
Cartlidge, D.P. 1976 *Construction design economics*. Hutchinson.
Cartlidge, D.P. 1973 *Cost planning and building economics*. Hutchinson.
Chalk, T. 1985 Double think on claims. *Building Technology and Management,* November, pp. 26 and 28.
Chapman, N.F.S.; Grandjean, C. 1991 *The construction industry and the European Community*. BSP Professional Books.
Chapman, K.; Karadelis, J. 1994 Building in the home of gods: an examination of the construction industry in Greece. *CIOB Construction Papers, No. 42.*
Chapman, N.F.S.; Grandjean, C. 1981 *The construction industry and the European community*. BSP Professional Books.
Chappell, D.; Powell-Smith, V. 1985 *JCT intermediate form of contract*. Architectural Press Legal Guides.
Chappell, D. 1991 Forms of contract – how to choose one. *The Architects Journal*, June, pp. 55-7.
Chevin, D. 1993 Design counsel. *Building*, November, pp. 36-7.
CIOB 1992 Inadequate tender documents and procedures; who pays? *CIOB Discussion Document*.
CIOB 1984 *Programmes in construction – a guide to good practice*. The Chartered Institute of Building.

CIOB 1975 The clerk of works and the site management team. *CIOB Site Management Information Service, No. 63.*

CIOB 1988 *Project management in building.* Chartered Institute of Building.

Clarke, R. 1990 Working with collateral warranties. *Chartered Quantity Surveyor*, August, pp. 19-20.

Codling, R.E. 1991 The contractor's accountant. *CIOB Technical Information Service, No. 133.*

Coles, E. 1990 Design manager: need training. *Chartered Builder*, December, pp. 16-17.

Collard, G. 1980 New deal for the client. *Building*, March, p. 55.

Cooke, B. 1981 *Contract planning and contractual procedures.* The Macmillan Press.

Cooke, B.; Walker, G. 1994 *European construction – procedures and techniques.* The Macmillan Press.

Cornes, D. 1993 Going for broke. *New Builder*, January, p. 14.

Cornick, T.; Osborn, K. 1994 A study of the contractor's quantity surveying practice during the construction process. *Construction Management and Economics*, **12**, pp. 107-11.

Cottrel, G.P.; Freeman, I.L. 1979 Quality in the construction of buildings. *Building Technology and Management,* November, pp. 36-9.

Cottrell, G.P. 1979/80 The builder's quantity surveyor. *CIOB Surveying Information Service, No. 1.*

Cowen, C. 1984 In delay there lies no plenty. *Chartered Quantity Surveyor*, January, p. 213.

Crawshaw, D.T. 1979 Project information at the pre-construction stage. *BRE Paper IP 27/79.*

Crowter, H. 1989 Head office overheads. *Chartered Quantity Surveyor*, February, pp. 28-9.

Davies, W.H. 1982 *Construction site production checkbook 4.* Butterworth Scientific.

Davies, C. 1984 The BPF system: The system in action. *Building,* June, pp. 31-3.

Davies, C. 1984 The BPF system: no going back. *Building*, June, pp. 28-9.

Dawson, H.C. 1988 A client's viewpoint. *Chartered Quantity Surveyor,* November, pp. 29-30.

Day, D.W.J. 1994 *Project management and control.* Macmillan.

Diggings, L. 1991 *Competitive tendering and European communities.* AMA.

Dixon, G. 1991 Finding a real alternative. *Chartered Quantity Surveyor*, February, p. 20.

DoE 1978 *Value for money in local authority house building programmes.* Department of the Environment.

Douglas, R.W.R.; Robertson, G.F. 1978 'Claim' is not a four letter word. *Building Technology and Management*, April, pp. 8-11.

Drayton, S.; Potts, K. 1992 Design and build in civil engineering. *Civil Engineering Surveyor*, February, pp. 4-5.

Drewin, F.J. 1982 *Construction productivity.* Elsevier.

Elliot, T. 1994 Surety bondsmen told to pay up. *Building*, March, p. 30.

Elliot, R.F. 1988 *Building contract litigation.* Longman.

Entwistle, M. 1990 Welcome alternatives. *New Builder*, November, p. 16.

Farrow, T. 1991 Acceleration; the agreement. *Chartered Quantity Surveyor*, September, pp. 24-5.

Fellow, R.F. 1989 The management of risk. *CIOB Technical Information Service, No. 111.*

Fellow, R. *et al.* 1983 *Construction management in practice.* Longman.

Ferry, D.J.; Brandon, P.S. 1994 *Cost planning of buildings.* Granada.

Ferry, D. 1991 *The organisation of procurement.* Thomas Telford.

Fleet, T. 1992 Legal argument – protecting against employer insolvency. *Building*, November, p. 34.

Fletcher, A.L. 1981 Attendance on sub-contractors. *CIOB Surveying Information Service, No. 6.*

Forster, G. 1978 *Building organisation and procedures.* Longman.

Fraley, A. 1990 Disputes procedure – mediation and the professional. *Building Technology and Management*, March/April, pp. 7-8.

Fraley, A. 1990 Mediation and the professional. *Chartered Builder*, March/April, pp. 7-8.

Franks, J. 1992 Design and build tendering – do we need a code of practice? *Chartered Builder*, June, pp. 8-10.

Franks, J. 1984 *Building procurement systems.* CIOB.

Franks, J. 1991 *Building contract administration and practice.* Batsford and CIOB.

Fryer, B.G. 1979 Management development – development in the construction industry. *Building Technology and Management*, May, pp. 16-17.

Furnston, M.P. 1986 *Law of contract.* Butterworth.

Geraghty, P.J. 1991 Environmental assessment and the developer. *CIOB Technical Information Service, No. 139.*

Gilbert, J. 1979 Defective work; the quality and standards expected is the contractor's responsibility qualified. *Building Technology and Management*, June, pp. 16-19.

Goodacre, P. 1979 *Delays in construction; ascertaining the cost.* CIOB.

Gordon, M.C. 1994 Choosing appropriate construction contracting methods. *Journal of Construction Engineering*, **120** (1), pp. 196-210.

Graves, F.C. 1982 The quantity surveyor and project management. *Quantity Surveyor*, February, pp. 28-32.

Gray, C. 1981 Management and the construction process. *Building Technology and Management*, March, pp. 18-21.

Green, R. 1986 *The architect's guide to running a job.* Architectural Press Management Guides.

Griffith, A. 1989 Design–build: procurement and buildability. *CIOB Technical Information Service, No. 112.*

Griffith, A. 1987 Quality assurance in building construction. *Building Technology and Management*, June/July, pp. 10-15.

Griffith, A. 1992 *Small building works management.* The Macmillan Press.

Groak, S.; Householder, J. 1992 Contractor's uncertainties and client's intervention. *Habitat International*, **4** (2/3), pp. 119-25.

Gruneberg, S.; Weight, D. 1990 *Feasibility studies in construction.* Mitchell.

Hambidge, B.W. 1982 Productivity, time and cost. *CIOB Technical Information Service, No. 10.*

Hamburger, D. 1992 Project kick-off; getting the project off on the right foot, *Project Management Institute* (*PMI*) **10** (2), pp. 115–22. Butterworth–Heinemann.

Hamilton, B. 1990 A team effort. *New Builder*, July, p. 18.

Hammond, D. 1993 Developing the art of project management. *Building Technology and Management*, March, pp. 2-3.

Harding, C. 1991 Without the fuss it would not be the same... I hope. *Building*, March, p. 23.

Harding, C. 1992 QSs should perform as they demand performance. *Building*, July, p. 32.

Hardy, T.J. 1992 Germany: challenge for the estimator. *CIOB Construction Papers, No. 11.*

Harper, D.R. 1990 *Building: the process and the product.* Chartered Institute of Building

Harper, W.M. 1977 *Cost accountancy.* M&E Handbooks.

Harrison, H.W.; Keeble, E.J. 1983 Performance specification for whole building. *Report on BRE Studies, 1974-1982.* Building Research Establishment.

Hawkins, R.A.P. 1993 Insolvency under JCT contracts. *CIOB Construction Papers, No. 24.*
Hayes, R. 1985 The risks of management contracting. *Chartered Quantity Surveyor*, December, pp. 197-98.
Hibberd, P.R. 1980 Interpretation of contract documents, the quantity surveyor's contribution. *The Quantity Surveyor*, April, pp. 60-1.
Hibberd, P.R. 1991 Certification. *CIOB Technical Information Service, No. 126.*
Hibberd, P.R. 1990 Contractor's design liability under the standard forms of building contract. *CIOB Technical Information Service, No. 118.*
Hillebrandt, P.; Andrews J. *et al.* 1974 *Project management: Proposals for change.* UCL–University College Environmental Research Group.
Hillebrandt, P.M. 1974 The capacity of the construction industry. *Building*, August, pp. 71-3.
Hillebrandt, P.M. 1977 *Economic theory and the construction industry*. Macmillan.
Hillebrandt, P.M. 1984 *Analysis of the British construction industry*. Macmillan.
Hillebrandt, P.M. 1978 Crises in construction. *Building Technology and Management*, March, pp. 4-6.
Hood, R. 1994 Design faults. *New Builder*, January, p. 14.
Horner, R.M.W. 1982 Productivity, the key to control. *CIOB Technical Information Service, No. 6.*
Houtte, V.V. 1991 *The impact of Europe upon the construction industry*. Lloyds of London Press.
Hughes, G.A.; Barber, J.N. 1992 *Building and civil engineering claims in perspective*. Longman.
Hughes, W.P. 1991 An analysis of the JCT design and build. *CIOB Construction Papers, No. 6.*
Hughes, W.P. 1992 An analysis of construction management contracts. *CIOB Technical Information Service, No. 135.*
Ingham, J. 1988 Meet in agreement. *Building*, September, p. 21.
James, H. 1994 Taking control. *New Builder*, February, p. 22.
James, M.F. 1994 *Construction law*. The Macmillan Press.
Jawad, A. 1994 Alternative dispute resolution – mediation as a cost effective method of dispute resolution. *The Structural Engineer*, **72** (7/5), April, pp. 109-12.
Johnson, R. 1983 Easy way out. *Building*, May, pp. 30-1.
Jones, E. 1993 Construction management and project management – the differences in structural and its impact on project participants. *International Construction Law Review*, pp. 348-65.
Joyce, R. 1995 *The CDM regulations explained*. Thomas Telford.
Kallo, G.G. 1991 Decision making in project management. *Civil Engineering Surveyor*, September, p. 31.
Keating, D. 1995 *Keating on building contracts*. Sweet & Maxwell.
Kilty, B.J. 1979 Is arbitration the answer? *Building Technology and Management*, November, pp. 30-3.
Kirk, T. 1982 Design under scrutiny. *Building*, June, pp. 30-1.
Knowles, R. 1984 A potshot at the BQ? *Chartered Quantity Surveyor*, March, p. 305.
Knowles, R. 1990 Adjudication is the answer. *Chartered Quantity Surveyor*, February, pp. 17 and 34.
Knowles, R. 1980 Contract completion after winding up. *Chartered Quantity Surveyor*, March, p. 221.
Knowles, R. 1986 Late issue of instructions. *Chartered Quantity Surveyor*, September, p. 9.
Knowles, R. 1984 BPF agreement – a reprieve for the BQ. *Chartered Quantity Surveyor*, August, p. 19.

Knowles, R. 1986 Project manager: the legal position. *Chartered Quantity Surveyor*, July, p. 11.

Knowles, R. 1985 Interest and finance charges. *Chartered Quantity Surveyor*, September, p. 73.

Knowles, R. 1980 Interest: direct loss and/or expense? *Chartered Quantity Surveyor*, August, p. 20.

Knowles, R. 1986 Time out. *Chartered Quantity Surveyor*, March, p. 9.

Knowles, R. 1985 Negotiation of claims. *Chartered Quantity Surveyor*, August, p. 22.

Knowles, R. 1994 Collateral warranties: are they worth the paper they are written on? *Chartered Surveyor Monthly*, February, p. 38.

Knowles, R. 1991 Time for change. *Chartered Quantity Surveyor*, September, pp. 8-9.

Kwakye, A.A. 1994 *Understanding tendering and estimating*. Gower.

Kwakye, A.A. 1991 Fast track construction. *CIOB Occasional Papers, No. 46.*

Kwakye, A.A. 1993 Alternative dispute resolution in construction. *CIOB Construction Papers, No. 21.*

Lai, H. 1989 Integrating total quality and buildability; a model for success in construction. *CIOB Technical Information Service, No. 109.*

Lean, W.; Goodall, B. 1977 Aspects of land economy. *The Estate Gazette.*

Lange, J.E.; Mills, D.Q. 1979 *The construction Industry*. Lexington Books.

Latchmore, A.; Foster, J.; Heaps J 1992 Warranties 2 – warranties in practice. *CIOB Construction Papers, No. 10.*

Lenard, D.J.; Roberts, C.F. 1992 The cost manager in the building procurement and delivery process. *The Building Economist*, June, pp. 18-21.

Lenard, D. 1992 Cost management in project delivery. *The Building Economist*, June, pp. 12-14.

Leong, C. 1991 Accountability and project management; a convergence of objectives. *International Journal of Project Management*, **9** (4), November, pp. 240-9.

Llewellyn, T. 1993 Why architects should plan for a tax–efficient design. *The Architects Journal*, November, pp. 28-9.

Lloyd, H. 1984 The BPF system: consultants architect's back the system. *Building*, June, p. 30.

Lloyd, H. 1990 It could have made me a fortune, your honour. *Building*, June, p. 1.

Lloyd, H. 1991 The year of retention draws to a close with a whimper. *Building*, November, p. 36.

Lloyd, H. 1991 Words not as good as a bond for thwarted contractors. *Building*, October, p. 34.

Lock, D. 1991 *Project management*. Gower.

Lord-Smith, P.J. 1994 *Avoiding claims in building contracts*. Butterworth Heinemann.

Lyall, D. 1984 Finance to fit the job. *Building*, July, pp. 27 and 29.

Mace, D. 1990 Problems of programming in the building industry. *Chartered Builder*, March/April, pp. 4-5.

Major, W.T. 1974 *The law of contract*. M&E Handbooks.

Mann, L.C. 1982 Insolvency; how to get the work finished. *Construction Surveyor*, July, pp. 18-21.

Manson, K. 1990 Extra payment to complete a building contract. *Clerk of Works*, November, p. 10.

Marshal, J. 1993 Construction industry – its ailments and cures. *Chartered Builder*, June, p. 5.

Masterman, J.W.E. 1992 *An introduction to building procurement systems*. E&FN Spon.

McIntyre, J. 1991 Disputes under review. *Chartered Quantity Surveyor*, November, pp. 16-17.

Mearns, I. 1987 *Fundamentals of cost and management accounting*. Pitman.
Mildred, R.H. 1982 Arbitration as applied to the construction industry. *Quantity Surveyor*, May, pp. 78-80.
Milne, M. 1993 Contracts under seal and performance bonds. *CIOB Construction Papers, No. 16.*
Minogue, A. 1992 How to minimise your exposure to off site goods. *Building*, January, pp. 32-3.
Minogue, A. 1991 The name is bond but the protection is limited. *Building*, October, p. 29.
Minogue, A. 1991 Choose whatever protection will best cover your risk. *Building*, November, p. 35.
Minogue, A.; Klien, R. 1984 Clash points. *Building*, April, pp. 28-9.
Minogue, A. 1992 Don't just blame design and build. *Building*, June, pp. 32-3.
Morris, A. 1994 Demanding cash up front. *Building Economist*, February, p. 20.
Morris, A. 1991 Insolvency; the practical issues. *Chartered Quantity Surveyor*, April, p. 8.
Morris, R. 1990 How they get the money: funding agreements. *The Architects Journal*, October, pp. 55-60.
Mosey, D. 1992 Collateral warranties – bridging the gap. *Construction Law*, October/November, pp. 131-4.
Moxley, R. 1993 *Building management by professionals*. Butterworth.
Murdoch, J.; Hughes, W. 1992 *Construction contracts – law and management*. E&FN Spon.
Murphy, N.; Chapman, S. *et al.* 1980 Image of the industry. *Building*, May, pp. 31-4.
Newell, M. 1977 An introduction to the economics of urban land use. *The Estate Gazette*.
Newlove, J. 1982 Variation by drawing issue. *Building Technology and Management*, September, pp. 21, 22 and 28.
Newlove, J. 1979 The issue of construction information. *Building Technology and Management*, July/August, pp. 2-5.
Nisbet, J. 1979 Post contract cost control: a sadly neglected skill. *Chartered Quantity Surveyor*, January, pp. 24-8.
NJCC 1983 *Code of procedure for two stage selective tendering*. NJCC.
NJCC 1991 *Code of procedure for single stage selective tendering*. NJCC.
Norris, C.; Perry J.; Simon P. 1992 *Project risk analysis and management*. The Association of Project Managers.
O'Hanlon, L.; Minogue, A. 1990 A genuine hold-up or is someone stealing time? *Building*, June, p. 42.
O'Reilley, J.J.N. 1987 *Better briefing means better building*. BRE.
Pain, J. 1993 Design and build compared with traditional contracts. *The Architects Journal*, November, pp. 34-5.
Pain, J.; Blyth, A. 1993 Are you using the right contract. *The Architects Journal*, October, pp. 27-8.
Palmer, A. 1989 Construction design cost optimisation. *Chartered Quantity Surveyor*, February, pp. 34-5.
Papworth, J. 1994 On-demand bonds. *Chartered Surveyor Monthly*, January, p. 44.
Parker, D. 1994 Safety catch. *New Builder*, July, p. 16.
Patterson, N. 1991 What do we know about the effective project manager. *Project Manager*, **9** (2), May, pp. 99-104.
Patterson, J. 1977 *Information methods for design and construction*. John Wiley & Sons.
Payne, A. 1990 Problems within building industry. *Chartered Builder*, September/October, pp. 33-4.

Peachey, B.J. 1979 Construction project control. *Chartered Quantity Surveyor*, January, pp. 10-13.

Penington, I. 1983 *A guide to the BPF system and contract*. The Chartered Institute of Building.

Pepperell, N. 1993 What a standard guarantee. *The Architects Journal*, May, p. 28.

Perry, J.G.; Hayes, W. 1986 Risk management for project managers. *Building Technology and Management*, August/September, pp. 8-11.

Petterson, D. 1992 Updating the profession. *The Architects Journal*, May, pp. 10-11.

Phipps, M. 1991 Is there any good reason not to certify on time. *Building*, June, p. 32.

Pike, A. 1994 The complete works. *Building*, October, p. 3A.

Pilcher, R. 1975 *Principles of construction management*. McGraw-Hill.

Piling, S.J. 1993 Brief formulation: the architect as manager. *International Journal of Architectural Management Practice and Research*, **1**, pp. 25-32.

Popper, P.A. 1994 Clear intentions. *New Builder*, July, p. 18.

Powell, J.M. 1992 Certificates and Payments: The new way forward. *Constructional Law*, pp. 13-17.

Powell-Smith V. 1990 *Problems in construction claims*. BSP Professional Books.

Powell-Smith, V.; Sims, J. 1985 *Building contract claims*. Collins.

Powell-Smith, V.; Sims, J. 1989 *Construction arbitration – a practical guide*. Legal Studies Services.

Powell-Smith, V.; Chappel, D. 1989 *Building contract dictionary*. The Architectural Press.

Prior, J. 1982 Weather and the Construction Industry. *Quantity Surveyor*, January, pp. 7-8.

Pugh, D.S.; Hickson, D.J. & Hinnings C.R. 1980 *Writers on organisation*. Penguin.

Ramsey, V. 1985 Get noticed. *Building Technology and Management*, pp. 26 and 29.

Ramus, J.W. 1993 *Contract practice for quantity surveyor*. B.H Newness.

Reiss, G. 1992 *Projet management demystified*, E&FN Spon.

Revay, S.G. 1993 Can construction claims be avoided. *Building Research and Information*, **21** (1).

RICS 1987 Contractor's direct loss and/or expense. *RICS Practice Pamphlets, No. 7.*

RICS 1984 Insolvencies. *RICS Quantity Surveyors Practice Pamphlets, No. 5.*

Rideout, G. 1983 On-demand bonds, licence to kill. *Building*, March, pp. 20-7.

Ridout, G. 1989 Construction management forum: reading lessons. *Building*, August, pp. 46-50.

Ritz, G.J. 1994 *Construction project management*. McGraw-Hill.

Roberts, D. 1984 Dispute avoidance. *Construction Surveyor*, January, p. 16.

Robinson, T.H. 1977 *Establishing the validity of contractual Claims*. CIOB.

Roskrow, B. 1992 Who needs lawyers. *Construction News*, January, p. 13.

Rougvie, A. 1988 *Project evaluation and development*. Mitchell-CIOB.

Rouse, D.J. 1986 Interim payments. *Chartered Quantity Surveyor*, October, p. 27.

Royce, N. 1980 Settlement of disputes. *Building Technology and Management*, January, pp. 13 and 15.

Rutter, D.K. 1993 Construction economics; is there such a thing? *CIOB Construction Papers, No. 18.*

Samuels, J. 1990 Buildability – an ongoing problem. *Chartered Builder*, September/October, pp. 14-16.

Saville, R. 1990 Getting the best from design-build. *Chartered Quantity Surveyor*, August, p. 9.

Saville, R. 1984 The BPF system: cause for concern. *Building*, June, p. 29.

Seel, C. 1984 *Contractual procedures for building students*. Holt Rinehart Winston.

Seeley, I.H. 1975 *Building economics*. Macmillan.
Seeley, I.H 1987 *Building quantities explained*. Macmillan.
Seeley, I.H 1978 Conference on avoiding a dispute. *The Quantity Surveyor*, September, pp. 246-8.
Seiderer, H.C. 1991 Arbitration v ADR. *Building*, February, p. 20.
Senior, G. 1990 Risk and uncertainty in lump sum Contracts. *CIOB Technical Information Service, No. 113.*
Severn, K.; Campbell, P.L. 1994 ADR; why treat the symptom, not the cause? *The Structural Engineer*, **72** (7/5), April, pp. 112-14.
Slavid, R. 1993 How the construction industry sees the architect's role. *The Architects Journal*, May, pp. 23 and 27.
Smith, L. 1984 A US spotlight on the BPF. *Chartered Quantity Surveyor*, May, p. 381.
Snowdon, M. 1980 Project management. *The Quantity Surveyor*, January, pp. 2-4.
Spring, M. 1993 Revealing design. *Building*, March, p. 35.
Staniforth, A.; Thompson, P. 1992 Collateral Warranties 1 – Introduction. *CIOB Construction Papers, No. 9.*
Stewart, A. 1994 Building on trust. *Building*, January, pp. 14-15.
Stephenson, D.A. 1982 *Arbitration practice in construction contracts*. E&FN Spon.
Taylor, N.P. 1991 *Development site evaluation*. Macmillan.
The Aqua Group 1989 *Pre-contract practice for the building team*. Blackwell Scientific Publications.
The Aqua Group 1982 *Contract administration for architects and quantity surveyors*. Granada.
The Association of Project Managers 1992 *Risk analysis and management*. The Association of Project Managers.
Thomas, R. 1992 *Construction contract claims*. Macmillan.
Thompson, P. 1991 The client's role in the project management. *Project Management*, **9** (2), May, pp. 90-2.
Thompson, N 1993 Building new role models. *Construction News*, December, p. 14.
Titmus, P.D. 1982 Design and build in practice. *Building Technology and Management*, April, pp. 9-12.
Toakley, A.R. 1990/1991 The nature of risk and uncertainty in the building procurement process. *Australian Institute of Building Papers, No. 4.*
Trickey, G. 1993 *Time is money*. David Langdon & Everest.
Trickey, G. 1983 *The presentation and settlement of contractor's claims*. E&FN Spon.
Trotter, P. 1982 The site manager. *Building Technology and Management*, November, p. 7.
Turner, D.F. 1977 *Building contracts: A practical guide*. George Godwin.
Turner, D.F. 1983 *Quantity surveying practice and administration*. George Godwin.
Turner, D.F. 1989 *Building contract disputes their avoidance and resolution*. Longman.
Turner, A. 1990 *Building procurement*. Macmillan.
Uff, J. 1991 *Construction law*. Sweet & Maxwell.
Uler, T. 1992 Risk management in construction. *The Chartered Builder (Australia)*, February, pp. 21, 23-4.
Vann, J.C. 1979 Bonds and guarantees. *Building Technology and Management*, May, pp. 18-19.
Vann, R. 1993 Contaminated land – digging up the dirt. *Chartered Quantity Surveyor*, October, pp. 13-14.
Wainwright, W.H.; Wood, A.A.B. 1979 *Variation and final account procedures*. Hutchinson.
Walker, A. 1982 Let's get organised. *Building*, June, p. 55.
Walker, A. 1984 *Project management in construction*. Granada.

Wallace, D.I.N. 1973 *Further building and engineering standard forms*. Sweet & Maxwell.

Ward, S.C.; Chapman, C.B.; Curtis, B. 1991 On the allocation of risk in construction projects *International Journal of Project Management*, **9** (3), August.

Ward, S.C.; Chapman, C.B. 1991 Extending the use of risk analysis in project management. *Project Management*, **9** (2), May.

Waters, B. 1983 The BPF system. *Building*, December, pp. 25-30.

Wearne, S.H. 1992 Contract administration and project risks. *International Journal of Project Management*, **10** (1), February.

Webb, C. 1991 Negotiating the management minefield. *Architect, Builder, Contractor and Developer*, April, pp. 18-20.

Welford, S. 1993 Total cost management. *Chartered Quantity Surveyor*, July, pp. 458-9.

Whitfield, J. 1994 *Conflicts in construction – avoiding managing, resolving*. Macmillan.

Willis, C.J.; Ashworth, A. 1990 *Practice and procedure for the quantity surveyor*. BSP Professional Surveyor.

Wood, D. 1993 Energy efficiency – healthy drift. *New Builder Special Features*, September, p. 34.

Woodland, D.; Chan, S.; Smith, N.; Merna, A. 1992 Risk management of boot projects. *Project Management*, March.

Wreghitt, J.D. 1989 Claims for prolongation and disruption. *The Surveying Technician*, October/November, pp. 8-9.

Yates, A. 1986 Assessing uncertainty. *Chartered Quantity Surveyor*, November, p. 27.

Yates, A. 1981 Old relations in new clothes. *Chartered Quantity Surveyor*, October, p. 64.

Young, B.A. 1993 A professional approach to tender presentations in the construction industry. *CIOB Construction Papers, No. 17.*

INDEX